Null Seeking Trials

Research Method for Surgery and Social Science

First published in Great Britain 2009
Reprinted with corrections 2011

Null Seeking Trials

Research Method for Surgery and Social Science

Patrick Mitchell

Brailsford

Table of Contents

Dedicated to Vic

Introduction

The dilemma of surgical research is well illustrated by a recent case considered for inclusion in a trial that I was running. Decompressive craniectomy is a treatment for raised intracranial pressure occurring after head injury in children. This is a major operation to remove a large section of the skull to allow for brain swelling. In the trial (the SUDEN trial), a random process was used to decide whether to do the operation.

In the following case description, some of the details have been changed to maintain anonymity. A family had been walking in a local beauty spot. They had three children, two boys of 6 and 8 and a girl of 5. They knew the area well and the children had run ahead together. At one point was a concealed sheer drop of several meters where a stream had undercut its bank. The five-year-old girl fell down this drop and sustained a head injury. Her brothers witnessed the fall. One of the boys climbed down to her and the other ran back to the parents. The parents arrived at the scene and found their daughter to be unconscious. The location was not directly accessible by road and so the helicopter air ambulance was dispatched. The crew found the child to be unconscious though breathing and airlifted her to the local neurosurgery centre where she was admitted under my care.

We used a device to measure the pressure inside the skull. She was treated using conventional measures. To begin with the pressure was in the normal range but as hours passed brain swelling caused the pressure to rise. Standard treatment was not completely successful at reducing the pressure.

This somewhat clinical and detached description is typical of medical case reports and does not comment on the fears and anxieties of the child's family. I had a daughter of a similar age to the patient. Their daughter had been a happy active child that morning and was now unconscious and in intensive care. We, the surgical team, did not know whether she would make a full recovery, die or be permanently brain damaged.

Neither did we know if this operation would help. We did know that it would involve a large incision, and removal of a large part of her skull, then weeks later – if she survived - doing another large operation to replace the removed skull with a titanium plate, which in the longer term was likely to leave her with an asymmetric head shape and an unsightly sunken area behind her eye. Furthermore, there was some suspicion among neurosurgeons that the operation's main effect was to prevent death in people who are the most severely injured, who then survive as seriously disabled, "vegetables" in popular parlance. Many patients who have the operation are saved, only to spend years in a coma on a hospital ward, their families repeatedly at their bedside hoping for recovery, but finally recognising that their loved one should have been allowed to die. Yet there was still the chance that the operation might reduce the girl's brain damage and improve her recovery. We just did not know where the balance lay.

In designing the trial (some time before this incidence), I had weighed the available evidence and identified the kind of cases we should enrol. This girl fitted perfectly. I had designed the trial in an untidy office, poring over reams of scientific literature, computer screens and research ethics guidelines. I was actually going to be doing the trial in a counselling room, with a family like mine, with a daughter like mine, whose life hung by a thread just twenty feet away. One question dominated the discussion: what would I do if she were my daughter?

I went to the bedside with the father, and spent several minutes in silence watching her. I carried out the kind of examinations doctors do, looking in her eyes, studying the monitors, but I was not paying any heed to the results – I was buying time, to think. Would I let my registrar cut *my* daughter open? Would I cut her open myself?

It took some time, but then I made a decision – the right one as it turned out. We did not randomize her and we did not operate. The pressure crisis passed. Her rehabilitation took months and she lost a year of schooling but she made a full recovery with her skull in tact.

Although it was my trial and I was motivated to recruit, when it came to it I could not. I could have persuaded the family to participate but that would have meant betraying their reliance on my judgement and replacing it with a computer's.

The null seeking trial is a research technique that was developed in response both to the difficulties of designing, setting up and running randomized trials in surgery and equally to the dilemmas facing surgeons and patients when considering participation in such trials. The specific trials that prompted its development were the above mentionsed SUDEN trial (SUrgical DEcompression for Neuroprotection) and the Surgical Trials of IntraCerebral Haemorrhage (STICH). STICH is a research programme run from Newcastle upon Tyne in England which I joined in 2002 when the first trial was recruiting. The STICH trials are randomized trials comparing surgical removal of brain haemorrhages with non-surgical treatment.

When I joined the STICH team there was little consensus amongst neurosurgeons around the world about when the operation should be used. There was more consistency of opinion within individual countries but wide variation between countries. Surgeons in the UK tended to be conservative, operating less than those in other countries. They would operate and remove haemorrhages under a fairly small range of indications. The patients they operated on tended to be quite young and with large haemorrhages in parts of the brain that are relatively non-functional but which were causing compression of other areas. Most people who have brain haemorrhages are older and the haemorrhages are most commonly in central areas of the brain. These were rarely operated on in the UK. In Japan the approach was very different. Japanese surgeons had a much stronger belief in the value of removal of haemorrhages and considered surgery on a greater proportion of patients. The STICH trials are international collaborations and the first one was characterised by relatively low recruitment within the UK compared to other countries.

As well as discussing the STICH trials with other neurosurgeons we also recruited patients ourselves and had many discussions with patients and families about the process of randomization. Brain haemorrhage is a severe form of stroke and many patients have a reduced conscious level so are unable to participate in discussions. Most families considered it important to participate in research for the benefit of others and agreed to randomization.

Collaboration with the STICH trial by other surgeons within the UK was limited. We discussed the trial issue with many surgeons who did the haemorrhage operation and two issues were recurring as reasons for non-collaboration. First was no “equipoise”. Second was the belief that the trial result would not help in practice. Equipoise is a legal term whereby the available evidence on either side of a question or issue is balanced so that it cannot be resolved one way or the other. If there is "no equipoise" about whether to operate, it means the available evidence is *not* balanced, so the question *can* be resolved. In clinical terms evidence is “balanced” when for alternative treatments, the argument in favour of one is no better or worse than the argument in favour of the other. In many individual cases there is no equipoise but to refuse participation in a trial because of no equipoise implies that a clinician never sees patients who are in equipoise. For many treatments this is unusual. If it is accepted that there are some patients who require surgery and some who do not and that there is a continuous range of situations in between, then this range *must* pass through an area of equipoise. The second objection is not so easy to argue against. Not many trials have been done in surgery and those that have often had little influence on surgeons' decision making. The reasons for this are outlined in Chapter 2 and they are cogent.

The null seeking trial is a research technique that was

developed in response to these problems. It differs from standard randomized trial methodology significantly in purpose and so also in design. The purpose of a standard trial is to observe a difference between the trial populations and infer that this difference is caused by the treatment - so called *causal inference*. The purpose of a null seeking trial, however, is not to make a causal inference about treatments, but to find the circumstances in which two treatments have the same effect.

The null seeking trial is still under development. To date no real null seeking trial has been conducted. Various simulations have been run aimed at assessing the method's feasibility but it has not yet been applied. This is because the objective and methods are unfamiliar to most researchers and the kind of research projects that null seeking trials are aimed at are large and expensive collaborative efforts spanning several years.

The purpose of this book is to provide the necessary background information for the design and conduct of a null seeking trial. This involves a certain amount of mathematics. Some prior knowledge of statistical and mathematical methods is assumed, at a level appropriate to the general scientific reader.

The mathematical techniques used here are primarily those of regression and in particular non-parametric regression. A brief introduction to this subject is included.

The null seeking method can be applied to a wide variety of research questions in different disciplines. This book concentrates on its application to the field of surgery. This is simply because it was developed by surgeons in order to solve problems encountered when applying randomized methodology to surgery. I make brief mention of the kind of research question that could be addressed in non-surgical disciplines. Most of the examples cited are from neurosurgery because that it is the subspecialist areas in which the methods were developed.

Chapter 1

Why Randomize?

The purpose of mathematical modelling is to apply the flexibility and reliability of mathematical reasoning to questions that are not immediately mathematical. Familiar examples such as Newtonian mechanics take experimental data and describe them with mathematical equations that are models of the real world. The real world is observed in experiments and the models allow us to interpolate and extrapolate beyond the experimental results themselves, to predict behaviour in situations that have not been directly observed.

The reliability of the predictions of models ranges from perfect (so far) for quantum mechanics, to very good for others models of theoretical physics. Those intended to model processes that are highly complex and incompletely understood are at the other end of the spectrum, where their predictions are often doubtful. These are the models used for situations such as disease processes, social behaviour, education and psephology (the branch of political science that deals with the study and statistical analysis of elections).

Processes with an element of unpredictability are modelled using random numbers and functions of them. The study of such models is the realm of statistics. In the early years of the 20th century statistics was successfully used to study processes like agricultural production, fermentation, and manufacturing. A typical problem might be whether a certain fertilizer, A, increased the cabbage yield per acre. Researchers accounted for unpredictable variations in the weights of individual cabbages by taking a large number of cabbages treated with fertilizer A, called the *experimental* group, and a large number that were not treated, called the *control* group, and comparing their average weights. They developed simple methods to guide judgements as to whether the difference observed in the average weights of cabbages was due to the effect of fertilizer A or was simply due to the unpredictability of individual cabbage weights. These basic methods were applied very successfully and are still widely used today, but as they were extended into more and more areas they began to fail.

With basic methods of comparison like these, if the numbers of cabbages are large, then even quite small differences in the average weights are highly unlikely to arise by chance. This makes the model quite sensitive to differences between "populations" of cabbages – but there is a problem. The model's main weakness is that there is no way of telling what the differences are *caused* by. If we compare the cabbage yield of two fields, one of which has been treated with fertilizer A, we may find a difference in the cabbage weights, but this may be because of a different amount of sunlight or irrigation or other factors that differ between the two fields, rather than because of the fertilizer. Getting round this problem to try to arrive at a reliable prediction or *causal inference* of the effect of some specific factor or "treatment", has been a major area of statistical research.

One way around the cabbage and field issue is to divide the fields into several blocks and add fertilizer A to alternate blocks interspersed between blocks with no fertilizer. The fertilized cabbages would then be mixed within each field so that any differences in other factors would largely cancel out — hopefully.

The problem of isolating the effects of a treatment from other causes has been greatest in the field of medicine. Early attempts to apply statistical techniques developed in agriculture and industrial production to medical problems failed because of unexpected effects. To compare two treatments of a particular disease, an obvious approach is to take one group of patients receiving one kind of treatment and compare them with another group receiving another kind of treatment (or no treatment). For any new treatment you can note the behaviour of the disease before and after the treatment is introduced. This is the still-widely used method of *historical controls* and is fraught with difficulty because the nature of many diseases changes over time as does the accuracy of diagnosis, the susceptibility of populations to the disease and other treatments available. It is impossible to tell whether a difference in the behaviour of the disease at different times is due to a particular treatment effect, or to one or more other factors. The historical controls method is a special case of a general principle, whereby any systematic association between treatment allocation and other influences on outcome makes it impossible to distinguish between them. Such associations cause *allocation bias*. Allocation bias occurs when there is a systematic difference between the patients receiving the two treatments, for example if one group is older on average, or has more severe disease, or contains a higher proportion of females etc.

An improvement on the historical controls method, aimed at avoiding allocation bias, is to take two alternative treatments, standard and new, for example, and a series of patients with a specific disease. The first patient is given the standard treatment, the second receives the new treatment, the third the standard treatment, and so on. This is a promising method but when put into practice it failed. It turned out that doctors often had existing beliefs about which treatment would be more effective. A patient who was improving spontaneously or responding to standard treatment would be more likely to be enrolled in the trial if the next treatment was standard than if it was new. Similarly, a patient who was deteriorating on standard treatment would more likely be enrolled if the next treatment was new. On average this means that patients receiving standard treatment tended to be improving but those receiving new treatment tended to be deteriorating, so had more severe disease. This type of behaviour, where clinicians and patients go to considerable lengths that have the unintended effect of thwarting a research project, seems perverse. Nevertheless it was a major problem in early medical research and was only overcome by imposing extreme measures. To see why it was such a problem we need to consider the research unit consisting of the individual patient and their treating clinicians.

A patient with an illness is concerned with their health. They have limited knowledge and experience of their condition and must put their faith in the judgement of any clinician they consult. Clinicians by their calling and training put the patient's interest above all else - it is hoped! Though they may be participating in a research project, that is of secondary importance. If they perceive an advantage to the patient from giving a certain treatment then they are bound to advise the patient accordingly. It would be unethical and professionally improper to withhold any such perceptions. Clinicians are well educated in the field and are likely to have wide experience of the behaviour of the illness and its treatments. If they are participating in a research project on a particular treatment, they can not help noticing whether patients on the new treatment appear to do worse or better than other patients. If they do notice such a difference, then conduct of the research project is a lower priority than securing what they perceive as the best treatment for their patients. This then leads to a conflict of interest between the aims of the research project and the clinicians' clinical judgement and, usually, clinical judgement wins.

From the point of view of research, clinicians have proved so resourceful at creating allocation bias that the only way around it has been to make it impossible for patients and clinicians to interfere with treatment allocation. This is achieved by using a system of allocation that they cannot predict. The necessary and sufficient condition for allocation to be unpredictable is that it is made at random. Randomization not only defeats clinician-mediated allocation bias but it also prevents all other forms of allocation bias, whether they are known about or not.

This necessary and sufficient condition means that if you want to avoid allocation bias you *have* to randomize. There is no other way. This fact has made randomization a standard in medical research. In practice it is not always feasible, however, and where it is not, evidence is unreliable because of potential allocation bias. One purpose of null seeking trials is to make randomization feasible in areas where currently it is not.

The nature of random allocation means that it is still possible to be unlucky with the randomization, leading to the phenomenon of *accidental allocation bias*. This occurs if the randomization process happens to allocate more poor-prognosis cases to one treatment than to the other. There are widely used methods for minimizing the chance of accidental bias, including block randomization and minimization. These methods work by arranging for equal numbers of particular types of patient to be randomized to each treatment. Block randomization is often used to ensure that individual participating hospitals or centres randomize equal numbers to each group. Allocations are arranged in blocks and each block has the same number of allocations for either treatment.

With minimization, the probability of randomization one way or the other is altered in the light of previous allocations. These two methods are effective at reducing the likelihood of accidental bias that concerns prognostic factors that are known, but they have no effect on the chance of accidental bias that concerns factors that are unknown. The only way of reducing accidental bias from such unknown factors is to increase the sample size.

From the point of view of the patient and clinician, making a decision about treatment at random, without the input of patient preference or professional judgement, is a significant imposition. The primacy of the patient's best interest over the interests of a research project is a fundamental principle that must be safeguarded. The issue of when to allow randomization is one that requires careful independent scrutiny, which is why all clinical research needs the approval of ethical committees that comprise clinicians, statisticians, patient representatives, ethicists and clerics.

The technical details of how to randomize have changed over the years. Some early trials involved tossing a coin. More commonly a table of random numbers would be consulted and one treatment would be allocated if the number was under 0.5 and the other if it was over 0.5. Nowadays computer-generated sequences of 1s and 0s are used.

Post-allocation bias

Another kind of bias occurs if there is any association between treatment allocation and the outcome of the condition other than via the direct effect of the treatment. This arises after allocation and it is called *post-allocation bias*. An important example is the *placebo effect*. Inert (dummy) tablets are known to have a beneficial effect on health by the power of suggestion. A trial comparing any tablet with no treatment at all is likely to find the tablet effective – even if there's nothing active in it. We are not interested in this effect as much as the chemical action of a particular medicine, so the placebo effect confounds our results. Randomization avoids allocation bias but it does *not* avoid post-allocation bias. There is no equivalent for post-allocation bias that is universally applicable and avoids bias of known and unknown causes. It is therefore addressed in different ways, depending on the specific situation, but remains a common problem and a reason to doubt the findings of a trial.

The solution to post-allocation bias resulting from the placebo effect is to ensure that neither the clinicians nor the patients know which treatment they are receiving. Methods of doing this are known as "blinding". Single blinding means the experimental subjects (or patients) do not know which treatment is being received and double blinding means that neither the subjects nor the experimenters (or clinicians) who measure and record their outcomes know. In practice this is usually done in drug trials were two batches of tablets are manufactured that look (and hopefully taste) identical but only one group contains the active drug. The placebo effect then acts equally in both trial groups and does not bias the difference between them.

In many situations it is not feasible either to carry out single- or double-blinding and in such cases the placebo effect is a genuine consideration but there are other possibly more potent causes of post-allocation bias. Pharmaceutical companies use trials to test, and to promote their drugs and products, and there are numerous ways in which post allocation bias can appear, particularly if there is a strong commercial incentive to find a particular result. Many involve compromising the control group. Placebos may not be as inert as they should be, perhaps being beneficial or toxic to the control group. A common problem arises with trials comparing two drugs where one is a standard drug (usually inexpensive) and the other is a new (expensive) alternative. The standard drug may be given at the wrong dose, or at the wrong time intervals, or in a form that makes it poorly absorbed or fail to reach the target tissue or disease process in some other way.

Null seeking trials do not avoid these problems of allocation and post-allocation bias but they modify their effect and may alter their importance when interpreting the trial.

Chapter 2

Trials

Randomization allows comparison between treatments without allocation bias. Currently such comparisons are primarily used in one way — to tell whether one or another treatment is better. Comparisons can be used in another way; in the field of engineering, particularly electronic engineering, comparison is used in combination with negative feedback, not to tell when things are different but to make them equal. Null seeking trials use the same combination of comparison and negative feedback to make things equal in the fields of medicine and social science. First we will look further into the conventional way of using randomization, to see why such an alternative may have its advantages.

A randomized trial is an experiment. It takes a population of patients with a particular illness and two treatments – say A and B. Patients are divided at random into two groups of similar size. One group is given treatment A and the other B. Measurements of the disease severity are then compared between the groups after the treatment. Outcomes such as the proportion who died or had strokes or heart attacks, or average blood pressures are commonly used. As patients come through the clinic door one after another it is usual to hand out a series of random allocations in turn.

The experiment of the randomized trial is interpreted with a mathematical model, which attempts to describe reality using a system of numbers and mathematical relationships. Mathematics is a highly reliable and predictable system. It is fairly easy to make the reasoning between assumptions and measurements, and the conclusions of the model totally watertight. The weak point in the process is the relationship between the model and the real-life situation. The randomized trial model matches the reality of the randomized trial experiment most accurately if post allocation bias is minimized. This means double-blinding with drug-containing and control tablets that appear and taste similar and have similar side effects so they cannot be distinguished after being taken (the control tablets may contain a placebo in a placebo-controlled trial or another drug in a comparative trial). Many questions in medicine fit this model well. Conventional randomized trials have been successful in dealing with such questions and have been carried out in their thousands. There are other areas, however, where these ideals cannot be met.

One problem area is the issue of concealing which treatment has been allocated. This may be because a drug has characteristic side effects that permit patients to recognize they are taking it. Or it may be because the treatments being studied are not tablets or drugs. If, for example, the treatment comprises psychological counselling or dietary education, then concealment is not possible. An intermediate situation occurs with surgical operations. Blinding patients can be feasible when comparing surgical procedures that use similar incisions because any gross differences in the incisions can be covered up using a suitably large dressing in all patients[1,2]. Often the most vexing problems in surgery, however, are not connected with comparing two operations but in deciding whether an operation should be done at all. The only way to blind patients in this situation is to perform so-called sham operations in the control group. This approach has been advocated [3]. One example is a trial that aimed to find out whether an operation to move the nasal septum (called septoplasty) helped people with obstructive sleep apnoea. Patients were randomly allocated to septoplasty or sham surgery. For the sham, the incisions for the operation were made but the septum was not moved [4]. Generally such a scheme is only regarded as ethically acceptable if the surgical approach is minimally invasive, which it is for procedures like septoplasty, endoscopy [5] and needle puncture [6]. More often, sham operations would carry significant risks, with none of the potential benefits of the real ones, so they are rarely ethically feasible.

Equipoise and neutrality

We are naturally wary when important decisions about medical treatment are made at random rather than being made by an experienced health professional. It is only considered ethical to use random methods if a condition known as *equipoise* exists. Equipoise in this context means that in the minds of both the professionals and the patient, the two treatments between which they are being randomized have equivalent efficacy according to the best currently available information. This is the definition of equipoise that surgeons use but another definition is used by ethical committees (see Chapter 4). An ethical prerequisite of randomization is equipoise about the relative merits of the treatments to be compared. This is a consideration in all trials but is not a major worry in many medical trials where the treatments may be limited in their effect or have recently been introduced so that their performance is not well documented in evidence.

In other situations, equipoise is a major issue. If there is good evidence that a treatment is effective then it is unethical to withhold it at random. For example, it would be highly unethical to withhold antibiotics from a child suspected of having meningitis, therefore a randomized comparison of antibiotics with no antibiotics cannot be done. Similarly it would be unethical to perform a major operation on a patient if it was felt to have little or no benefit.

The need for equipoise has various consequences in the conduct of clinical trials. In cases where equipoise is a major issue, there is a tendency to recruit patients where the clinician and patient both believe the treatments to be similar. Consequently if they are right in this judgement (and they frequently are because clinicians have extensive experience of the behaviour of treatments), then the trial tends to be neutral, showing neither one nor the other treatment to be better. In cases where neither treatment can be said to be better, the problem of proving a negative prevents us from concluding that they are equivalent, so these trials have little value in guiding treatment decisions.

Crossovers and the Hypothesis Problem

When treatment cannot be concealed, post-allocation bias from placebo-like effects can arise, but perhaps more importantly another problem arises – that of crossovers. Crossovers are patients who have been randomly allocated to one treatment but for whatever reason end up having the other – that is, they have "crossed over". This frequently happens in surgical trials which may have crossover rates as high as 45% in one or other treatment group[7]. The recently completed STICH trial[8] of surgical removal of brain haemorrhages, found 26% of patients who had been randomized to the conservative (non-surgical) group went on to have surgery while 6% who had been randomized to surgery did not receive it. The discrepancy arose because while it is possible to operate at any time after a patient has been randomized to non-operative treatment, once an operation has been done following randomization to surgery it cannot be undone!

The "hypothesis problem" is a consequence of crossovers. The trouble with crossovers is that they are selected to cross from one treatment group to the other after randomization has taken place. This means that the reason they have been selected to crossover is liable to introduce post-allocation bias. An example of this was seen in the STICH trial [8]. Many surgeons around the world believe that surgical removal of brain haemorrhages is helpful when patients are lapsing into coma. The patients who did not receive surgery and who went on to deteriorate towards coma often had the operation anyway. This meant that the group of patients who received surgery contained some who were selected from the non-surgical group because they were doing badly, and this introduced a post-allocation bias against the surgical group. Getting round this bias is quite difficult. It appears most obviously if you simply compare the patients who had surgery with those who did not (so-called *treatment received* analysis), but even if you eliminate all the crossovers and compare those who were randomized to have surgery and had it, with those who were randomized not to and did not in a so-called *per protocol* analysis, then bias is still present because the deteriorating cases from the no-surgery group have been withdrawn. The way to analyze the results and avoid this form of post-allocation bias is on an *intention to treat* basis. With this kind of analysis you compare all the patients randomized to surgery (whether they had it or not) with those randomized to no surgery (whether they had it or not).

In order to accommodate the ethical requirement that crossovers are allowed, the hypothesis has to be changed from the clinically relevant question to one that can be tested with an intention to treat analysis. The clinically relevant question is of the form *Is this treatment better than that one?* or *Is this treatment better than no treatment at improving the outcome?* The one that can be tested is more like *Is an intention to treat with surgery associated with a better outcome?* or *Is a policy of early surgery associated with a better outcome?*

Such a hypothesis, if stated fully, would be: *An intention to treat with surgery – where various factors beyond the control of the investigators may prevent treatment being given as randomized and the investigators may decide to change the treatment on the basis of a lack of continuing equipoise in the light of evolution of the condition or otherwise*. This makes the hypothesis ill defined and specific only to the investigating surgeons who will have been selected to participate in the trial, and to the historical time period in which it was conducted. It contains a component of clinical discretion that is likely to change in the light of the trial result. Applying the result of the trial involves two dubious assumptions: first, that the practice of the participating surgeons during the trial is representative of surgeons at large and, second, that it is representative of their practice in the light of the trial result! This is the hypothesis problem.

How damaging the hypothesis problem is depends on the

crossover rate, the result of the trial and the perceptions of clinicians. Results from intention to treat studies are generally applied by a clinician by approximating their own policy with the triallists'. Whatever the triallists' crossover policy, the approximation is better with lower crossover rates. If the triallists' cross over policy was similar to their own, they accept the result with a higher crossover rate than otherwise. A problem is that the triallists' crossover policy is generally neither consistent nor published, so we do not know what it was. Also, the result is more credible if it is either neutral or highly significant. If it is marginal, the post-trial clinicians will be less tolerant of crossovers. These factors cannot be quantified but the neutral STICH result with 26% crossovers was accepted, the neutral SPORT result with 45% was not.

These two problems, trials frequently being neutral and answering questions not directly applicable to individual patients, bedevil research in many fields including surgery. Trials are costly, taking thousands of man-hours and several years to complete. The intention is to gain new information so that we can improve the treatments on offer. As a surgeon I will change the advice I give and the techniques I use if reliable evidence comes along to suggest I can improve my results. Too frequently however, when results of trials become available, they are either neutral or not directly applicable to my practice, or in some other way not compelling. The resources expended on such trials are in part wasted. Neutrality and applicability are the most important reasons why so few trials in surgery are done and why those that are done often do not gain wide support. They also prevent a majority of research proposals getting off the ground in the first place.

Chapter 3

Multiple Binary Trials

The randomized trial model is designed to address research questions that have yes or no answers – *binary* questions. Sometimes research questions in real life also have yes or no answers, but such situations are in a minority. More commonly we have to adapt the research question we are interested in to make it into a yes or no question in order to allow use of the randomized trial model. As an example consider the question *What is the effect of aspirin on the risk of having a heart attack?* This question potentially has a complex answer. Aspirin reduces the risk of heart attack by a quantifiable amount and this varies depending on the risk factors and clinical condition of the individual patient. In order to conduct a randomized trial we have to derive from this complex situation a yes or no question. It is likely to be of the form: *In this group of patients, does aspirin reduce the incidence of heart attack by more than 10%?* This process of casting a research issue as a yes or no question is common to virtually all randomized trials and it has significant implications for how research topics are tackled. Trials give more information than a simple yes or no answer. A large amount of data is generated. It allows secondary issues such as how much benefit a drug has, and in whom, to be addressed. The primary research question – and the one that is most reliably answered – is the yes or no one. Supplementary data may be useful but only as qualifying information to the primary trial result, not as information that stands independent of the primary result. Some large trials are designed with enough power to answer several yes or no questions. The carotid endarterectomy trials discussed below are examples, but these are in effect multiple simultaneous trials that still have a basic binary design.

In a favourable situation the question we are interested in can be approximated closely by a yes or no question. In the case of aspirin, the question is simply: *Should this patient be prescribed aspirin to reduce their risk of heart attack?* Details such as the amount by which the risk is reduced are of lesser importance. It is reasonable to pick the threshold at which we would consider prescribing aspirin; based on costs, risks and convenience; then conduct a trial that asks whether aspirin reduces the heart attack risk by more than this threshold. Several large trials have been conducted on this particular question and show that aspirin does reduce the risk [9]. Trials aimed at testing drugs often have this advantage of fairly close approximation between the trial's binary question and the clinically important issue.

The situation is not so simple when testing treatments that are more dangerous than aspirin, with less clear advantages. Examples are the many cancer chemotherapeutic drugs or treatments that are not drugs. In these cases the issue of interest may only be crudely approximated by a yes or no question. Consider, for example, the condition of low back pain or lumbago, a common and disabling condition that is associated with arthritis of the spine. It is notoriously difficult to treat. Literally hundreds of treatments are available but the evidence that any of them has a lasting benefit is slim indeed. Research in the area has been hampered because the condition is a complex one with multiple contributing factors. We may want to research the effects of a particular treatment, but approximating the question *What is the effect of this treatment in lumbar back pain?* with a yes or no question such as *Does this treatment improve lumbar back pain by more than 10%?* is so crude as to be of minimal use.

One operation that is used for lumbar back pain is spinal fusion. In this operation, screws are driven into adjacent vertebral bodies and connected together with a system of rods to fix them in relation to each other. Additional measures are taken to encourage the two vertebral bodies to join together into one bone. The idea behind this operation is that the pain comes from joints, not bones. If the joint movement is eliminated by fusing the bones together the joint pain should stop. The problem is that the operation is a major undertaking, with considerable risk. If we want to find out if this operation is effective in relieving lumbar pain, we plan a randomized trial. We have to come up with a yes or no (binary) research question that is meaningful when applied to patients after the trial has been done and achieves equipoise, so it is ethical and feasible to address.

Lumbar fusion is quite widely done. Many spinal surgeons have great experience with the procedure and the results that it gives, and have a good idea of who to use it on. Opinions do differ, but all surgeons agree there are some patients with back pain who should not have the operation. Patients with arthritis throughout the lumbar spine are one such group. There are other groups of patients that surgeons agree should have the operation. Patients with a slip of one vertebra on another (spondylolysthesis) are such a group. In both of these groups, equipoise does not exist so they are excluded from trials. For other patients, with one or two arthritic disc joints and no slip, surgeons are often in equipoise, so several trials have been conducted on this group[10-13]. The results of these trials were similar to each other. Around 20% of cases crossed over from one treatment to the other. The intention to treat analysis showed some improvement of back pain with surgery, but the effect seemed minimal [14]. Few surgeons changed their practice in the light of these results.

Non-binary models

Outside of the subject of randomized trials, binary models are not the only ones in use. If we classify models according to the nature of their output variable, trials are binary because their output is either yes or no. Other kinds of variable are *categorical*, *ordinal categorical* and *continuous*. Categorical variables are things like types of rocks, animals or treatment. We can place such things in categories without implying a specific relationship between the categories. Ordinal categorical variables are those in which the categories are strictly ranked. The number of children in a family is an ordinal categorical variable. The categories are 0, 1, 2, and so on,and there is a strict relationship between them. Two is less than 3 and more than 1. Some variables seem to be ordinal but are not. Coma scales are one example. The commonly used Glasgow Coma Scale (GCS) gives a measurement of the depth of coma, ranging from 3 for profoundly unconscious to 15 for fully conscious. The scale is made up by adding three subscales, relating to limb movements (from 1 for no response to 2 and 3 for primitive responses to pain, up to 6 for normal movements in response to instructions), eye opening (from 1 to 4) and speech (from 1 to 5). The total score is a number from 3 to 15, which appears to be ordinal but is not. A patient with a score of 9 is not strictly more awake than one with a score of 8 because there are several ways in which scores of 8 and 9 can be made up. Even if each subscale is ordinal (which is doubtful), a person with a GCS of 9 could have a lower score in one or two subscales than someone with a score of 8 and therefore cannot strictly be more conscious. A simpler measure of consciousness would be whether the person responds to pain, verbal command, neither, or both. If each response is given a 1 or 0 and the results are added, then the resulting scale of 0–2 is ordinal. One is strictly more responsive than 0 and strictly less responsive than 2.

Continuous variables are those that can take any value over a particular range. They are the most useful and allow for the most powerful statistics. Mean blood pressure is an example; others are age, weight, body mass index, and blood sodium concentration. The number of children in a family is not continuous because you can't have half a child, or any other fraction of 1.

Any of the kinds of variables discussed here can be approximated by binary numbers. This means that research questions of interest that cannot be reasonably addressed with single binary trials can be reasonably approximated with multiple binary trials.

This effect of approximating a question relating to a continuous variable with multiple trials that yield binary results can be seen in the case of surgery for carotid stenosis. The main supply of blood to the brain arrives via the left and right internal carotid arteries. These arteries commonly become affected by atheromatous disease, the same process that causes heart attacks. The atheroma narrows the carotid arteries and makes the lining irregular and prone to provoke the formation of small clots. These small clots are liable to detach from the atheromatous plaques and travel in the blood stream where they lodge in the small arteries that feed the brain or eyes. This mechanism is one of several causes of stroke and mini-stroke.

When someone has a stroke or signs of a mini-stroke, the doctors seek to find out the cause. One of the tests involves scanning the internal carotid arteries to see if they are narrow; if they are, it is possible to remove the narrowing with an operation called a carotid endarterectomy. While this operation may prevent strokes in the future, it carries a small risk of causing a stroke itself. This situation led to the research question: *Which patients, if any, benefit from endarterectomy, and specifically how high should the risk of stroke be to justify the risks of surgery?*

A number of randomized trials have been conducted on this subject and collectively they largely answer the question. The North American Symptomatic Carotid Endarterectomy Trial (NASCET)[15] ran from 1987 to 1996 and recruited 2226 patients and the European Carotid Surgery Trial (ECST)[16] ran from 1981 to 1994 and recruited 3024 patients. The two trials were large enough to show the effect of surgery in several subgroups and they had similar findings. Patients with more than 50% stenosis benefited, and those with more than 90% benefited the most. Patients with stenosis of 30–50% did not show any difference and those with less than 30% showed a detrimental effect [17].

The feasibility of conducting these large trials is unusual in surgery and was down to this several factors. Stroke is managed by physicians who have typically been involved with several trials on medicines. Because the effects of surgery are not seen immediately (it takes some years before the risks are balanced by the stroke-prevention effect), individual surgeons do not have a clear idea of benefits from their own experience, making them more inclined to equipoise.

The carotid endarterectomy situation illustrates a more general principle about maladies that affect individuals to different extents, and treatments that carry risks of their own. The argument in favour of such a treatment, technically known as the *indication* for the treatment, cannot adequately be modelled as a binary variable, but it can adequately be modelled as a continuous variable. If the indication for a treatment is modelled as a continuous variable, in general it can range from a negative value, through zero, to positive. This equates to patients in whom the treatment is clearly not indicated (at the negative end of the range of the continuous variable) to patients where it is clearly indicated (at the positive end). I will refer to the range of this variable as the *spectrum of indication* and the variable itself as an *index of indication.*

The trials on carotid endarterectomy used the percentage of stenosis as the index of indication and gave multiple binary results that approximated it as an ordinal categorical variable. The categories were >90%, 50–90%, 30–50% and <30%. The indication for surgery is a function of the index, being strongest in the >90% category, persuasive in the 50–90% category, in equipoise in the 30–50% and negative in the <30%. The raw data generated by the trial can be used to approximate indication as a function of the continuous variable percentage stenosis. The trials were not primarily designed to do this but they demonstrate the existence of this function, which is a central part of the null seeking trial.

Chapter 4

Equipoise

Most medical research has been conducted in a fair and reasonable fashion but there have been a small number of regrettable incidences in the past in which the subjects of medical research were not treated fairly. They may have been given no option but to participate as has happened to some prison inmates and military personnel. In some cases research subjects were not told that they were participating in research, or were not told about known risks of what was involved. In a study in which the effects of the disease syphilis were being observed, penicillin was withheld from patients for more than 20 years after it was known to be effective. This was because the researchers in the Tuskegee Syphilis Study wished to observe the behaviour of the untreated condition [18–20]. Episodes like these in which research subjects were not treated in a way we would expect to be treated ourselves led to regulation of medical research with the aim of preventing recurrences. Over several decades the current system of ethical committee approval was developed.

When considering randomized trials, ethical committees have to decide whether it is reasonable to choose at random between two different treatments. If current evidence suggests clearly in the minds of the medical profession that a particular treatment is the best one, then it would be unfair to withhold that treatment from people suffering from the disease, which means it would be unfair to randomize between the two treatments.

Medicine has borrowed the term *equipoise* from law to

describe the necessary condition for ethical randomization. In legal terms it relates to a situation in which the weight of evidence in favour of one side of an issue is equal to the weight of evidence in favour of the other. Freedmann introduced the term to medicine in 1987[21] and took it to mean a situation whereby each of two competing treatments has a reasonable body of responsible professional opinion to support it. If this condition exists, then ethical committees can approve a trial.

Equipoise defined in this way is a good enough condition for a trial to begin ethically, but not to bring it to a successful conclusion in many fields. In the kind of drug trial mentioned in Chapter 2 if there are reasonable bodies of professional opinion for and against the use of the drug in a particular circumstance it means that professionals on either side of the issue will probably be able to agree that they cannot be confident in their opinion. The evidence available to them on the behaviour of drugs, particularly newly introduced drugs, is virtually all from published research which is accessible to everyone. Furthermore, there is no difference between the drug administered by one person and by another, so there is no problem in accepting that the conclusions of one professional's published results are applicable to another's practice. As you might expect, however, things are not always so simple.

One field in which the concept of equipoise applied by ethical committees is not enough to bring a randomized trial to a successful conclusion is surgery. A surgical operation is a dramatic life event for a patient. It carries risks of death or permanent disability as well as the potential for immediate and lasting relief of symptoms or prolongation of life. The patient often has no option but to place faith in the skill and judgement of the surgeon. The surgeon takes seriously the formation of their recommendations and judgement, and gives the patient relevant information. In addition, surgical operations are difficult to do, and outcomes depend on how well they are performed; and surgeons do the operations themselves. In general, to be competent at an operation surgeons have to have done it many times and they cannot help but gain experience in how their patients do after the operation as they personally perform it. Much of the existing evidence used by surgeons comes from their own personal experience and that of their teachers and mentors, rather than in published form[22] and equipoise varies between surgeons with differences in opinion and different skills. This gives them a body of information about how to counsel patients that is gained from direct personal experience. This is a very different situation from advising a patient whether or not to take a new drug, or an old drug with a new indication.

Patients deciding whether or not to have an operation are counselled by individual surgeons. When they come to their decision, one factor that usually carries little weight is what other surgeons would or would not recommend. Patients are interested primarily in what their surgeon recommends, and the reasons for it. That being so, if a patient and surgeon are considering randomization then the ethical committee principle of equipoise has little relevance because it does not refer to the individual surgeon's opinion but simply to there being a substantial and competent body of surgeons who recommend each option.

The condition under which patients are happy to be randomized is when they and their own surgeon discuss all the issues and available information and still cannot decide which treatment is best. This is the meaning of equipoise that I will adopt. There are two reasons for using the same term, despite the possible confusion with the way it is used by ethical committees. First, it is the same as the original legal meaning, in as much as the evidence does not lead those involved in a decision one way or the other. Second, it is what most surgeons mean by the term.

The difference between equipoise in medicine and surgery makes surgical equipoise specific to surgeon–patient combinations. Although there may be a lack of randomized trial data for a particular procedure, that will not leave surgeons with no opinion about its value, so it is difficult for equipoise to be defined by one surgeon for another, or by trial organizers for investigating surgeons. This means that it is difficult to get agreement between surgeons on what trials should be done and what the inclusion criteria should be. The scale of trials necessarily means that they have to be done by many clinicians in collaboration, and a lack of common equipoise makes it difficult to design them. Furthermore, trials have to be funded and that funding is decided on by research committees. The same lack of common equipoise makes it difficult to design trials in the first place, and also makes it difficult to achieve agreement on committees as to what should be investigated, making it difficult to fund surgical trials.

Another consequence of equipoise adds to the lack of support that surgical trials receive. Surgeons have equipoise and are prepared to randomize patients only when they feel the potential benefits of surgery are closely matched by the potential risks. They generally have extensive experience of the results of operations they do, so they are likely to be quite accurate in this assessment. If so, the trial will be neutral, or will show little effect, as is often seen in trials of surgical procedures[8, 23]. As a consequence, surgeons tend not to be motivated to join proposed trials because they think, often correctly, that the result will not change practice.

The null seeking trial was designed to address problems relating to consensus, support for trials, and recruitment. Its impact on other problems in surgical trials is discussed in Chapter 13. The principle of the null seeking trial could be applied to any alternative treatments. For simplicity I will assume a surgical versus medical treatment question, and refer to the treatments as S and M accordingly, but the conclusions can be generalized.

Chapter 5

Serial Binary Trials

The carotid endarterectomy trials discussed in Chapter 3 allow us to approximate the true relationship between an index of indication (percentage stenosis) and the benefits of surgery. When trial data are not available we cannot do this, but still there is a *perceived* relationship and spectrum of indication. Ask a surgeon which patients would benefit from an operation and they will have an opinion about it. Their opinion on who would benefit, who would be harmed, and who would on average not be affected is their perceived spectrum of indication and that differs from surgeon to surgeon. The goal of a research project is more closely to approximate the true spectrum of indication with the perceived ones. The ultimate goal is to find the true spectrum of indication, but the limitations of statistical methodology mean that we can never define it exactly. We can only approximate.

A hypothetical situation is illustrated in Figure 5.1. I have chosen the "true" relationships between the index of indication for surgery and outcome for two treatments, treatment S and treatment M. I have then modelled individual patient outcomes as a random function centred on this selected "true" relationship. These hypothetical functions are used to illustrate the relationship between them and the results of binary trials. In real life the true relationships are not known. What available evidence there is will yield areas of consensus of indication for treatment S, indication for treatment M, and between them a zone of equipoise.

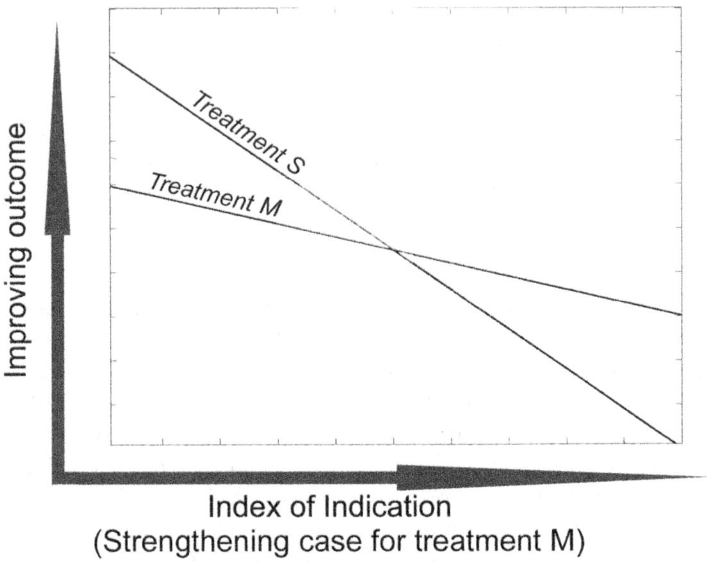

Figure 5.1 Hypothetical functions relating indication to outcome for two alternative treatments of the same condition. Treatment S could be surgical and treatment M medical, for example. If carotid endarterectomy were under study the index of indication could be the percentage stenosis and outcome could be the stroke rate.

Hypothetical zones of equipoise are added in Figure 5.2. We begin by considering a series of binary trials, which can only be performed within these zones. The position of the two lines is not known during the trial, but if the result of a binary trial reflects treatment efficacy accurately, then it will depend on the relationship between the crossing point (or *null point)* and the zone of equipoise. If the consensus is too optimistic for treatment S (Figure 5.2a) then the zone of equipoise lies in the area where treatment M is better, and so the results of a trial will favour treatment M. This result will weaken the case for treatment S, so that if a further trial is done the zone of equipoise is likely to have moved, perhaps as shown in Figure 5.2b. This time the trial would favour treatment S. Now, with both trial results in hand, if a third trial is done, the zone of equipoise will be squeezed so as to converge on the null point. If the zone of equipoise is symmetrical about the null point (Figure 5.2c) or the treatments are not different in their effect then the trial result is likely to be neutral. As treatment S is invasive or risky, enthusiasm for it will be lessened by a null result because patient opinion, economics and the medical Hippocratic principle of "first do no harm" all argue against the use of invasive measures that are ineffective or in which the benefits are balanced by risks. The result of a trial would be to move equipoise to the left because cases where treatment S was previously indicated would now be in equipoise.

A final situation to consider is that of Figure 5.3. Here, treatment S is never beneficial compared to treatment M. In this situation, a binary trial conducted in the zone of equipoise would result in a weaker case for treatment S and the zone for a second trial would move left, where again the case for S would be weakened and so on until equipoise was banished altogether and treatment S abandoned.

These examples illustrate two features of series of binary trials done on the same clinical question. First they allow the zone of equipoise to converge on the null point – if there is a null point. If there is no null point, series of trials allow equipoise progressively to move in favour of the better treatment, until the other is forced out altogether. This would happen even if at the beginning of a series of trials, the inferior treatment could not be tested in some patients for ethical reasons, because everyone believes it to be beneficial. Serial trials result in a shift in the zone of equipoise that is always towards the null point. This is a necessary and sufficient condition to establish negative feedback that operates on the zone of equipoise to centre it on the null point.

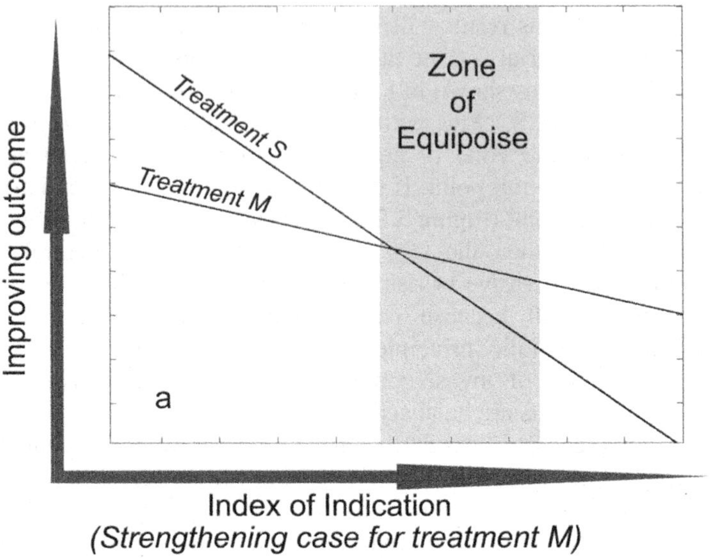

Figure 5.2. (above and facing page)
Three diagrams illustrating two alternative treatments, S and M, that might be compared in binary trials. At the right extreme are cases where the perception is that treatment M is clearly better. At the left extreme are cases where treatment S is perceived as being clearly better. Whether or not the perceptions are correct there will be a zone of equipoise (represented by the grey column). The trial is done within this zone. If the initial zone of equipoise lies in the benefit region for treatment M (a) the trial will show this benefit. Treatment M will be indicated in cases previously in equipoise and the zone of equipoise will move left. If the initial zone of equipoise lies to the left in the area of benefit for treatment S (b) the trial will move it right. If the initial zone of equipoise lies symmetrically over the null point (c) the trial will show no difference between A and B. If treatment S is more invasive, costly or unpleasant than treatment M it will not be used in these cases and the zone of equipoise will move left, towards the side of greater benefit for treatment S. The black lines show hypothetical true relationships that are shown for illustrative purposes. The corresponding real true relationships are not known during the conduct of trials.

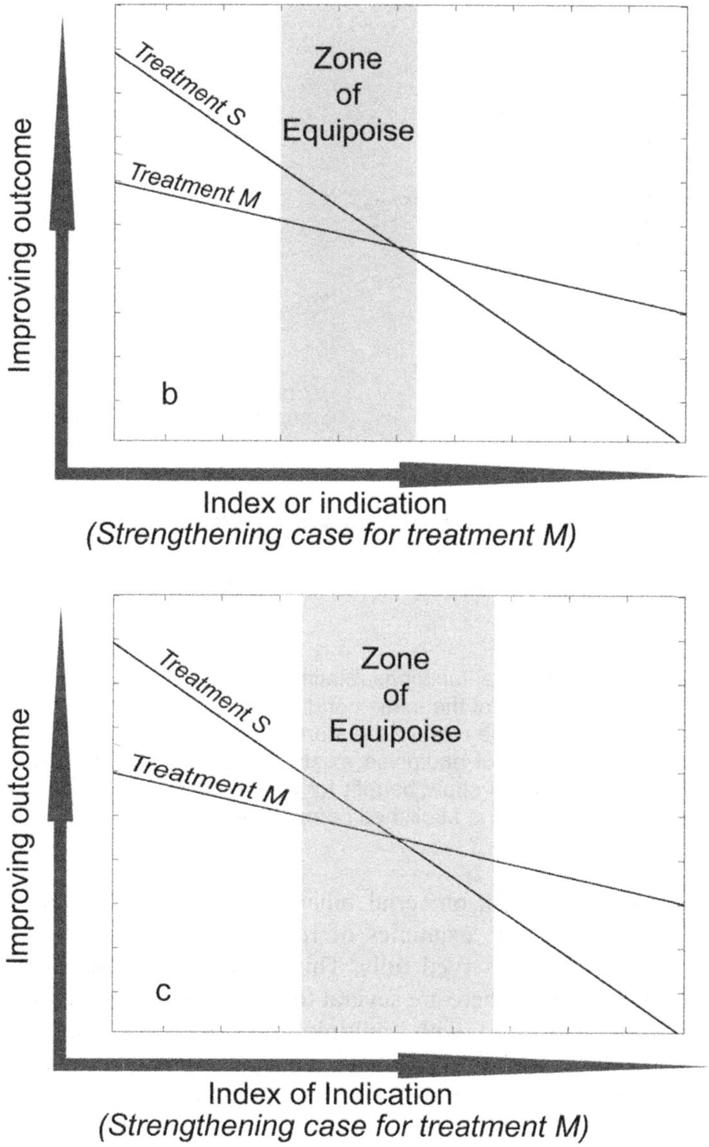
Zone
of
Equipoise
Treatment S
Treatment M
Improving outcome
b
Index or indication
(Strengthening case for treatment M)
Zone
of
Equipoise
Treatment S
Treatment M
Improving outcome
c
Index of Indication
(Strengthening case for treatment M)

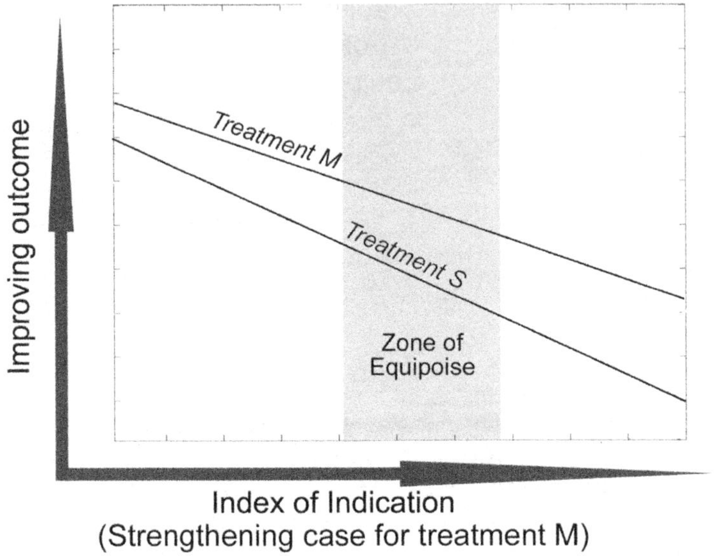

Figure 5.3. Hypothetical functions relating indication to outcome for two alternative treatments of the same condition where one treatment (S) is inferior and has no true role. At the start of the trial it may be felt to be effective and a zone of equipoise exists, but binary trials repeated in series will consistently show benefit for treatment M, moving the zone of equipoise left until it is abolished completely.

The above discussion on serial binary trials is a theoretical one, because there are no examples of research questions where this process has been observed fully. This does not mean the process would not happen. There are several topics in surgical research that have been addressed with multiple trials but, as with carotid endarterectomy, trials have run concurrently rather than in sequence. In order for trials to be truly serial, there would have to be enough time between the completion of one trial and the commencement of the design stage of the next for practice to change. Randomized trials take between 10 and 20 years from the initial design to the final result becoming available and changing practice takes several more years to occur after the results are published. Large-scale randomized trials in surgical specialities have only been underway in the last 30 years, so leaving insufficient time for even two truly serial trials to be completed.

Chapter 6

Null Seeking Trials

In Chapter 5, the principle of how a true spectrum of indication is approximated by a perceived spectrum of indication in the light of serial binary trials was discussed. A null seeking trial exploits the same principle of negative feedback, but it takes the process further. Instead of mounting serial binary trials where the perceived spectrum of indication changes in the light of the previous trial, and before the next one is designed, one single trial is carried out. The result is updated and made public as the outcome for each patient becomes available.

The purpose of a null seeking trial is not to show a significant difference between treatments but to allow convergence between equipoise as perceived by surgeon–patient combinations and a point where treatments are truly equivalent – the null point[1]. A system of negative feedback is used to do this.

A zone of equipoise within a spectrum of indication ranging from indication to contraindication is a more prominent feature in surgery than in medicine. This is because surgical operations have inescapable risks that are highly dependent on operator skill and coexisting morbidity, other than the disease being treated by the operation. For patients at high risk from surgery or anaesthesia, but with little to gain from the operation,

---

1 This is a simplistic first approximation. No-one would go through an operation with no net benefit! The real null point is when the patient feels that the discomfort and inconvenience of having an operation are exactly balanced by the gains.

surgery is not indicated. Conversely, for patients at low risk with much to gain from surgery, it is indicated. Whether these consensus are correct is an interesting question but not one that can be addressed using randomization because there is no equipoise. It will be assumed for now that they are. Between these extremes lies a spectrum of varying strengths of indication. At points on this spectrum, indication turns to equipoise and then to contraindication. Essential to the null seeking trial is making this spectrum explicit. The null seeking trial model is based on a quantifiable outcome, which is modelled as a function of a numerical index of the strength of the indication for surgery. Figure 6.1 is an example of the type of functions involved in the treatment of unruptured intracranial aneurysms. The outcome variable is remaining life expectancy in years, the index is simply age.

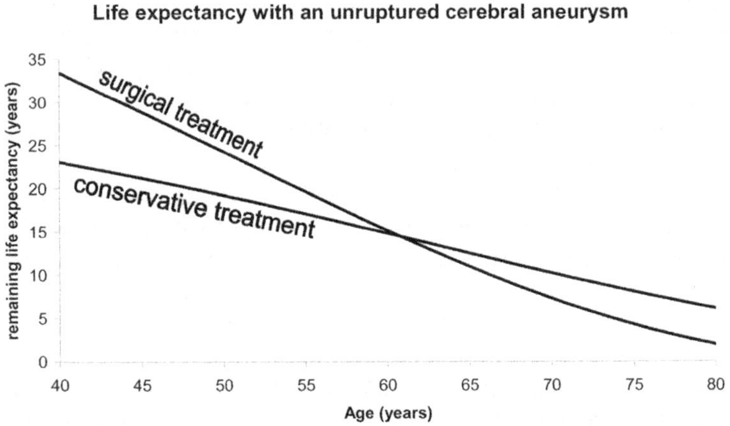

Figure 6.1. Plots of life expectancy for persons harbouring unruptured intracranial aneurysms of 13–24 mm in diameter, treated with or without surgical craniotomy and clipping (see references [24] and [25] for full details of the analysis). The clinical situation has properties that would allow a null seeking trial; both outcome (remaining life span) and an index of indication (age) can be expressed as numbers.

The curves represent estimates of the kind of functions whose crossing point the null seeking trial is designed to define.The idea of the null point as a zero crossing point is fine in theory but in practice the outcome measures used in trials are often fairly crude and do not accurately reflect the whole patient experience of the treatment and its outcome. The life expectancy example of Figure 6.1 is an example. It takes no account of the immediate stress, discomfort and inconvenience associated with undergoing a major brain operation, including the loss of a driving license for 6 to 12 months[2]! Most patients would not go through that for a theoretical week or two of life extension. A year or more and they might be tempted. This illustrates how in practice the null point may not be the crossing point.

Features can be added to this basic description to improve the utility of null seeking trials. One is that all patients treated could be included, rather than only those who are randomized. A coding system is used in Figure 6.2 to differentiate between patients who were selected to have a treatment and those who were randomized to it. This allows the randomized cases to be looked at in isolation avoiding allocation bias, or all cases to be looked at together helping to define the lines outside the zone of equipoise. The degree of allocation bias introduced by the selection of cases can be modelled and measured by comparing the lines with and without the non-randomized cases. For example, if the line plotted for treatment M with only randomized cases included is in a different position from that for treatment M plotted with all cases, or only non-randomized cases, it indicates that selection is occurring to a degree that is quantifiable by the difference between the lines. This is considered further in Chapter 8.

Practical implementation

The null seeking trial could be accessed via an Internet website. When a patient agrees to participate, their details (including factors

2 UK regulations

that influence the index of indication) would be entered. The investigating surgeon and patient could refer to the display of recent results of the trial. This is a plot of outcomes for other cases in the trial (on the y-axis) against the index of indication (on the x-axis).

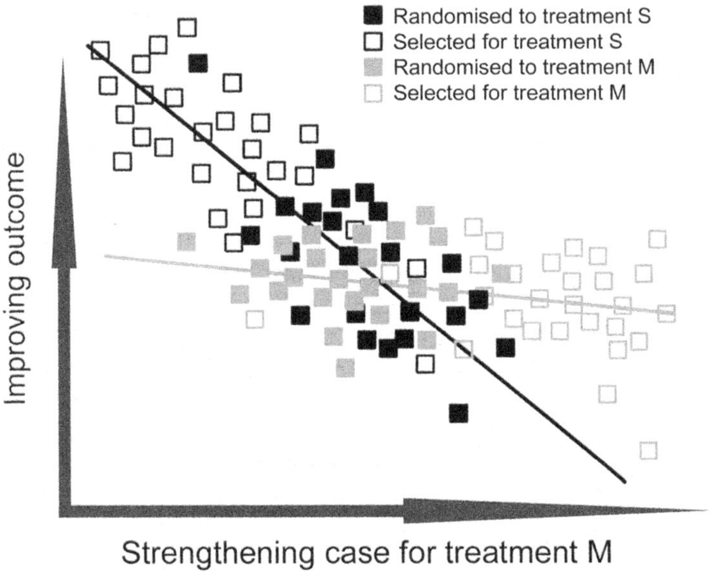

Figure 6.2 An illustration of what the data display in a null seeking trial might look like. Randomized cases are shown as solid squares and non-randomized as hollow squares. If the clinician's impression of which treatment is indicated is correct, we would expect to see a substantial non-randomized group of cases having treatment S towards the left-hand side of the graph and a substantial non-randomized group of cases having treatment M towards the right-hand side of the graph.

The value of the index for the patient being considered would be shown, and the displayed would include those data from patients who have been randomized and from patients who have not.

On the strength of the results viewed, the investigating surgeon and patient would decide whether to have treatment S or treatment M or to be randomized.

At the start of a null seeking trial, there would be a delay between the first patients being recruited and thier results becoming available. During this time it may be that few patients were randomized, with the picture first being built up from results obtained from non-randomized patients.

There is no need for the index of indication to be defined in the same way for all participating surgeons or investigators. It could be a function of variables including patient age, sex, their general condition and the severity of the disease being treated. As long as they are specified at the beginning of the trial (or *pre-specified),* several functions of one or more of these variables could be used as the index of indication — the multiple linear regression model for outcome prediction being an obvious one.

Investigators would be encouraged to select their preferred index and often to revisit the issue about what index to use. This diversity would benefit the trial because it would widen the range of individual indices over which randomization was done, making it easier to define the lines of Figures 6.2 to 6.7 as well as the null points. To illustrate this, let us consider a trial on surgery for sciatica secondary to lumbar disc disease. Surgeon 1 may use pain severity as his index of indication while surgeon 2 may use the duration of pain. It is likely that surgeon 1 will be in equipoise and randomize over a fairly narrow range of pain severities. On the other hand, as this surgeon is not using pain duration, he or she will randomize over a wide range of durations. The converse is true for surgeon 2. The result is that randomization will cover a larger range of both severity and duration than it would if only one index was used.

Neither is there any need for all investigators to use the same outcome measure as long as they are all pre-specified. The outcome measure used may vary from investigator to investigator and the same investigator may use different measures on different occasions. This property could be useful when investigating treatments with significant ethical dilemmas such as the treatment of pre-term infants or anyone with a serious head injury where lives may be saved at the cost of severe disability. Outcomes of survival and quality of life could be available. The use of decompressive craniectomy as a treatment for post-traumatic raised intracranial pressure is an example.

Decompressive Craniectomy for Brain Injury

After a severe head injury, the brain is prone to swelling. This swelling happens inside the skull, which allows little space for expansion, so it can lead to increased pressure in the head. This makes it difficult for blood to get to the brain which is then at risk of not having enough glucose and oxygen for normal functioning, as well as having to recover from the injury. Brain damage in addition to that caused directly by the injury may result, leading to worse permanent disability or death. Treatment is aimed at preventing this additional brain damage by controlling swelling and consequent raised pressure, so maintaining blood flow. Several medical treatments are available for this but they do not work for everyone and the pressure may still rise dangerously in some people despite their use. A surgical operation called decompressive craniectomy is available for these patients. In this operation a large piece of the skull is removed to create space into which the brain can expand, relieving the pressure. It is thought to save lives but many of the survivors have permanent and severe disability. A null seeking trial of the treatment could be launched in which the proportion of cases surviving was one outcome measure and the quality of life among survivors was another. Both of these could be plotted against an index of indication, perhaps the rise in intracranial pressure. We could make distinctions such as that between a low chance of survival but high quality of life among survivors and a high chance of survival but low quality of life among survivors. Such distinctions may inform our decision making. Using two outcomes would not only facilitate the conduct of the trial, but also make its result more valuable.

A different situation is where there are two related outcome measures, such as length of survival and length of good-quality survival in the treatment of malignant brain tumours. Using the two outcomes would enable situations with long poor-quality and short good-quality survival to be distinguished from those with short poor-quality and long good-quality survival.

Multiple options and post-hoc selection

Binary trials are designed to have a primary result, usually a single yes or no answer to a single question. They are also designed to give a number of secondary results which are intended to inform the interpretation of the primary result. Prominent among the secondary results is the concept of subgroups.

Subgroups are subsets of the data that are analyzed with particular secondary questions in mind. In the case of the STICH trial, the primary question was: *Does surgical clot removal improve the outcome after brain haemorrhage?* Among the secondary analysis were questions aimed at subgroups, such as: *Does the primary result hold for men or women or small clots or large clots?* Subgroup analysis can be informative but it is notorious for being often misinterpreted or subverted for arbitrary purposes.

The problem is a temptation to regard the subgroup results as having equal status to the primary result, which is far from the truth. If the primary result says the treatment is ineffective and a subgroup is found in which it appears to be effective, this is more likely to be misleading than true. There are only two circumstances when such a finding can be credited. They are when the effect seen in the subgroup is so extreme as to be highly improbable to arise by chance, and when the same effect is seen repeatedly in the same subgroup in several trials. Both of those situations are unusual.

Subgroups form the background to the primary result and

not alternatives to it. They should never be quoted in isolation from the primary result – but they often are[26].

The reason subgroups can be so misleading is that on the one hand they often give a simple and desired message that is a powerful tool for promoting sales or other agendas, and on the other hand the objection is subtle, counter-intuitive, and not widely understood. Key to the interpretation of a trial result is its *p* value. This is the probability of the result arising by chance assuming that the treatment has no effect. By convention we accept the result as a real difference if p is less than 1 in 20 (or 0.05). This means that if a treatment is totally ineffective, 1 out of 20 trials to test it would be expected on average to produce a p value at or below 0.05. If we wish to promote the treatment, and these 20 trials have been done, we would choose the trial that had produced the significant p value, and not mention the results of the others. Presented with this trial data, an audience is likely to accept that the treatment is effective. The most detailed examination of the trial's conduct would not show any cause for concern; the critical problem is the 19 trials that haven't been mentioned.

The same principle applies to multiple analyses of a single trial. If we come up with 20 variations on the basic hypothesis, we would expect to find one of them to have a significant p value, even if the treatment is totally ineffective. We could test to see its effectiveness in old people or in young people, in men or in women or, more subtly, we could use a range of different statistical techniques to calculate the *p* value. Eventually we would expect to get a p value that is significant. This regrettably common practice led the economist Ronald Coase to remark "If you torture the data long enough, [they] will confess". When we have our low p value we can then present the results of a trial and no examination of the trial's conduct or data can demonstrate that there is anything wrong with our conclusion. The vital piece of missing information is that we have tried numerous variations unsuccessfully before getting the p value we want. To avoid detection of this behaviour all we have to do is conceal the fact that we have done multiple analyses. What is more, people not schooled in statistics – but armed with powerful software packages – can do this to a data set without realising that it is misleading.

This problem is more general than trial subgroup analysis. It is referred to as post-hoc analysis and it takes many forms, all of which have in common the feature of selecting a hypothesis or analysis method in the light of the result it gives. This is why triallists now have to publish details of their planned analysis method *before* the analysis is done, generally before the randomization code is broken. It is my belief that triallists should also be required to place the raw data in the public domain within a specified time of the trial result being published, partly to enable others to check the result directly. This is not current practice.

Null seeking trials would not be immune to this problem. They would generate data sets just as susceptible to post-hoc analysis. Post-hoc selection of null points, indices of indication or zones of equipoise would be just as misleading as would hypotheses or analysis methods. Why then are investigators encouraged to select from several possible indices of indication or outcome variables?

The advantage of having several indices of indication for investigators to choose from, or allowing them to define their own, is that just as opinions differ over where equipoise lies, so they differ over the determinants of surgical indication. The reason for allowing the options is to make the trial as inclusive as possible. Another advantage is that using multiple indices increases the range of individual indices over which randomization is likely to occur.

There are various drawbacks to allowing so many options, however. In order to allow a choice of different indices of indication and outcome measures, the necessary data for each of them must be collected for every patient. The amount of data it is feasible to collect puts a practical limit on the number of options.

Another problem is post-hoc hypothesis selection. Although the purpose of the trial is to find the null point, the published result must put this in context and is of the form: *The indication for surgery depends on index A; these are the ranges of*

A where surgery is indicated, is not indicated and is in equipoise. Using several options leads to several results from which to choose. If towards the beginning of the trial, a non-pre-specified index shows good correlation with outcome or interaction with treatment then there is enough time in the remainder of the trial to test this index and for negative feedback to correct a pure post-hoc selection effect. This is not true for promising non-pre-specified indices that appear towards the end of the trial, unless the finding is important enough to justify extending the trial to give sufficient time for negative feedback to act. Analysis can compensate for multiple options and it is a lot easier to develop complex statistics to match the real world than it is to contrive experiments in the real world with the restrictions necessary to fit simple statistics. Compensation for multiple options costs significance – multiple analysis compensation always increases p-values.

The commonest analysis method for binary trials is to *dichotomize* the outcomes by dividing them into two categories, such as good and bad, and then to compare the *proportions* of good and bad outcomes in the treatment and control groups. This is an insensitive method giving higher p values than other methods on similar data. The normal method of comparing *variables* for their predictive value is with regression or analysis of variance (ANOVA), which can allow for multiple options. Related methods are used in null seeking trials. The resulting p values are higher than for single-option analysis of the same data but the sensitivity gained by not dichotomising means that low p values can be achieved even with multiple options on moderate-sized data sets. However, the options must be limited in number and pre-specified.

Null seeking trials may collect several variables relevant to the indication for surgery that could be included in a compound index, and more than one outcome measure, leading to numerous possible combinations. To allow reliable extrapolation of results beyond the trial after its conclusion, it is necessary to pre-specify a limited number of combinations of index and outcome as options. Combining multiple outcome measures where possible into single ordinal or quasi-ordinal scales helps with this. What is not so obvious is *when* to pre-specify the options. It is not necessary to do this at the beginning of the trial because negative feedback can correct early errors, but it should be done by about half way through the trial. This is likely to give enough time for important trends to emerge and to agree consensus on what options to pre-specify.

Treatment variability

One problem that affects binary trials is variable treatment performance. When the treatment is a complex one administered by people (such as a surgical operation or an education program) it will not be the same from one subject to the next. Worse is that there are likely to be systematic differences between how the treatment is administered by different people, hospitals, countries and cultures.

Only a small minority of surgeons generally participate in a trial, and treatments as they administer them may not be representative of those of their peers outside the trial, or even of themselves after the trial is completed. This variation impairs our ability to apply the results of trials. As an example, the leading proponents of a particular operation are often reluctant to participate in a multi-centre trial because they feel that in their own hands, equipoise lies further towards the more aggressive use of surgery than that proscribed by the trial. In this belief they may well be correct. Also by virtue of their interest and experience they are likely to be the best at administering the treatment under study. This means that a randomized trial may provide high-quality evidence of the efficacy of the treatment when administered with mediocre competence.

This situation has the danger of resulting in a neutral or negative trial that will be used to deter the best performers from administering the treatment, when the best interest of patients would be served by demonstrating that the treatment administered by the best performers was beneficial, so encouraging other surgeons to match their performance rather than discouraging all from administering the treatment.

A particular advantage of null seeking trials is their ability to address this problem. Differences between participating surgeons can be addressed by allowing investigators to view the trial result as a whole or in subgroups, specifically only those patients treated in their country, their unit or by themselves. This allows the best performers to participate, to randomize and to establish their own zone of equipoise. It has the disadvantage of creating numerous subgroups, with the attendant risks of post-hoc selection.

A complicating factor in this situation is that whether selection is post-hoc becomes dependent on viewpoint. From the viewpoint of an individual surgeon, his or her local results are pre-specified. From the point of view of other observers, with no particular reason to focus on one or other individual hospital or surgeon, selection of the apparently best performers is post-hoc. This means that while a trial is in progress no multiple option correction needs to be made for surgeons looking at their local results, but following the completion of the trial such corrections must be made for general application of the results. This is analogous to correcting for multiple subgroups in a binary trial.

Statistical analysis

The mathematical methods necessary for analysis are discussed in Chapters 8 to 11. Here I present a graphical description of what the emerging results of a null seeking trial could look like. This is based on a computer-simulated null seeking trial with some simplifying assumptions that would rarely be appropriate in real life. It is necessary to define the functions that the lines for treatments S and M represent separately from the treatments themselves. For this reason I have assigned the symbol Ψ to the function that relates the outcome to the index x for treatment S, and the symbol Φ for treatment M in Figures 6.2 to 6.7. In this simulation, both are straight-line relationships with data points uniformly distributed around them with constant variance.

The principal used for analysis is: for each point in the xy

plane of Figures 6.2 to 6.7 to calculate the probability that that point lies in between the treatment S and treatment M lines, rather than above both or below both. This probability is a function of (x,y) and is called Q. Illustrative results from such calculations are displayed as a three-dimensional plot in Figure 6.3, but for practical purposes a contour plot is easier to read accurately. In Figures 6.4 to 6.6 the contours labelled 0.025 indicate points with a probability of 0.025 of lying between the treatment S and treatment M lines.

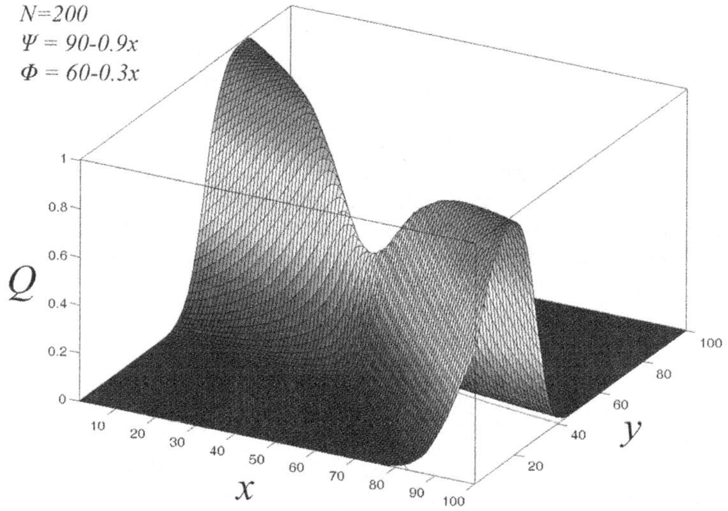

Figure 6.3 Three-dimensional plot of the probability (Q) of points in the *x-y plane* lying between the two treatment lines Ψ and Φ, using the functions of Figures 6.4 to 6.6. The crest corresponds to the space between the functions Ψ and Φ. At the point where they cross Q=1/2. This simulation is of a null seeking trial after 200 cases have been recruited.

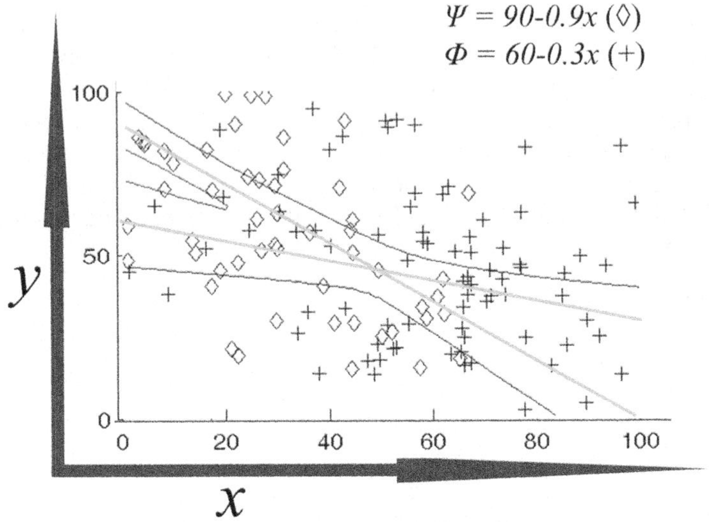

Figure 6.4 Central part of a simulation of a null seeking trial after 200 cases have been recruited, showing data points, least-squares lines (thick grey lines) and contour lines for Q=0.025 and Q=0.975 (fine black lines). Far from their crossing point, these approximate to 95% confidence limits for the two lines.

Any point further from the lines than this contour has a lower probability of lying between the two lines. Conversely the 0.975 contour indicates points with a probability of 0.975 of lying between the lines. The areas enclosed by the 0.975 contours represent points that have a probability greater 0.975 of lying between the two lines; in other words they are highly likely to lie between the lines. This in turn means that the two lines are likely to be separate at values of the index *x* where such points exist.

These values were chosen because far from the null point they approximate to the familiar 95% confidence limits of the position of the two lines. Both the confidence limits and the region where the lines are separate (inside the 0.975 contour) are of interest in the conduct of the trial.

The value of this scheme is that it gives an idea of when

the lines for treatments S and M are significantly separated. The expected interpretation would be that if there is at least one point with a high probability of lying between the two lines, then they are at significantly different places for that value of the index variable. Where there is no point with a substantial probability of lying between the two lines, they can be said not to be significantly different. In the area where the lines are not significantly different, equipoise may exist.

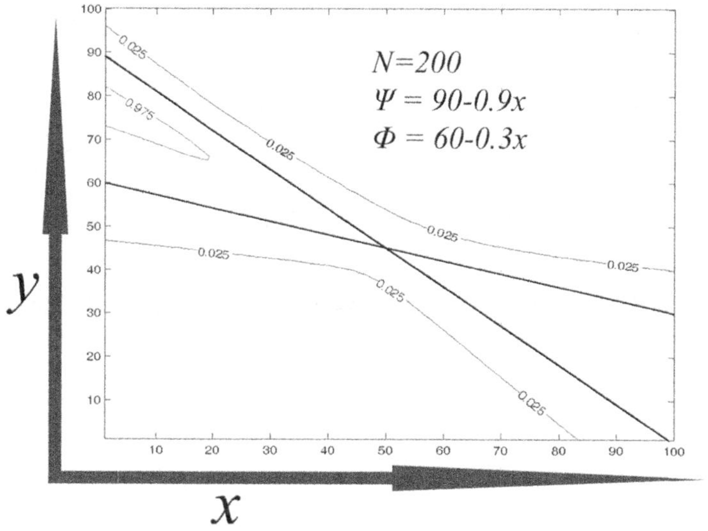

Figure 6.5 The same simulation as in Figure 6.4 with the data points removed and the crossover magnified for clarity. The 0.975 region appearing on the left-hand side indicates the emergence of a significant difference between Ψ and Φ and may inform the zone of equipoise.

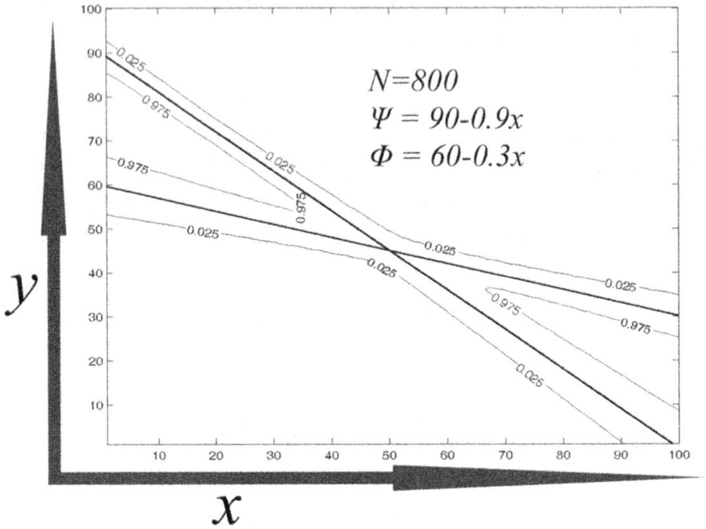

Figure 6.6 The same simulation as in Figures 6.4 and 6.5 after 800 patients have been recruited.

Chapter 7

Negative Feedback

Negative feedback is a central concept in fields such as electronic engineering, but it is a ubiquitous principle that applies to virtually any system that is stable. Boats float on water at a stable average depth because of negative feedback. Motors turn at a stable speed because of negative feedback. Objects remain on the ground and don't float up into the air or sink into the earth because of negative feedback. Electronic amplifiers produce faithful reproductions of recorded sound because of negative feedback. Cyclists balance because of negative feedback. Uni-cyclists also balance because of more negative feedback. Largely because of its importance in the field of electronic engineering, the theory of negative feedback has been quite highly developed.

Negative feedback systems have a stabilizing effect because they automatically correct any deviations from the stable situation that arise. In the case of a boat floating on water, the force supporting the boat is directly proportional to the volume of boat submerged below the waterline. If waves plunge the boat deeper into the water, more boat is under the water so the up-thrusting force increases and pushes the boat out again. Similarly if wave action pushes the boat out of the water, the up-thrusting force reduces so that it is less than the force exerted by gravity on the boat, which pushes it back down again.

The null seeking trial concept is designed around achieving negative feedback. It is the main reason why null seeking trials differ so much from conventional trials. Although negative feedback theory has been well developed, the null seeking trial is a different situation from that encountered in electronic engineering. For this reason we need to consider it from first principles.

There are three things that define the nature of negative feedback. First, how much there is of it. Second, how long the time delay is before it is applied. Third, how much "damping" there is in the system. In general, as long as the damping is adequate, the more negative feedback is applied, and the quicker it is applied the more stable the system will be.

Let us consider first how much negative feedback is applied. How does a car keep between the white lines defining a lane while driving down a motorway? This is achieved by negative feedback from the person who is driving the car. A driver who notices the car drifting to one side or the other of the lane, will correct it by steering in the opposite direction, bringing the car back to the middle. How accurately the car is driven in the middle of the lane is controlled by the three factors mentioned above. First, the amount of negative feedback. A driver who notices that the car is drifting left and turns sharply right will rapidly regain the middle of the lane. By turning right more gently, the driver will return to the middle of the lane more slowly. In technical terminology the relationship between the amount of steering by the driver and how far the car has drifted off the middle of the lane is referred to as the *gain* of the negative feedback loop.

Second, is the issue of the time delay. In this scenario it translates into the time it takes from the driver noticing that the car has drifted from the middle of the lane before their corrective intervention changes the car's direction. Clearly the faster this happens, the less the car will have drifted off and the more quickly it will return to the centre of the lane.

The third and final factor is damping. Damping applies a "drag" on the rate at which the car can move from one side of the lane to the other. Without damping, the driver, on noticing a deviation to the left would correct it by turning to the right, and pass the centre of the lane. The car would move towards the right edge and the driver would then apply further negative feedback input to steer the car back to the left and the same thing would happen. The car would oscillate between the sides of the lane and not stay in the middle. In this situation steering geometry and the driver's corrections achieve effective damping. The car does oscillate from side to side, but too little to notice so the consequences of inadequate damping are not familiar experiences for drivers. The rolling and pitching of boats are an example we are more familiar with.

Applying these concepts to the gathering of data in a randomized null seeking trial allows us to identify what features of the data are associated with these features of negative feedback.

Amount of feedback

As with the car in the motorway lane, the amount of feedback is determined by the scale of the response to a deviation from the stable state. In the case of a null seeking trial this means the amount by which opinion shifts in response to an emerging trial result that shows the difference in the performances of the two treatments. Suppose we are mounting a null seeking trial of surgical resection of brain tumours. After the trial has begun it shows that on average those patients randomized to have surgery are doing better than those randomized not to. Opinion is expected to shift towards a more aggressive surgical policy. Suppose in this case the spectrum of indication is a scale of 0 to 100 where current equipoise is in the 50–70 region, with 0–50 being rarely operated on and 70–100 being normally operated on. Suppose now that the survival advantage that the trial is showing for the 50–70 group is 4 months. If the trial has low-gain negative feedback this may lead to a change of opinion by perhaps 5 points on the indication scale. This would mean a shift in the randomization range to 45-65. If the trial had high-gain negative feedback, it might cause a 20-point shift, and randomization would move to the 30–50 region.

We can maximize the gain of the negative feedback loop by picking pre-specified indices of indication to correlate with surgical benefit as closely as possible.

Time delay

With the null seeking trial situation this can be quite substantial. We have to wait until results of the trial are available and in the above-mentioned example of brain tumours this means waiting to see how long patients survive – which could be quite a while. If we are studying a fairly aggressive form of brain tumour we might have an expected survival time of a year or so. That means an average of a year's delay before each individual result is known, and an individual result won't influence opinion significantly. It may take 6 months' worth of results before a convincing pattern emerges, which would lead to an 18-month time delay before applying feedback. With more benign types of brain tumour the average survival might be 6 to 7 years, leading to longer time delays.

The rate at which the result of the trial converges on the true null point is dependent on this time delay and this will be faster when the delay is shorter. Also a shorter time delay implies that less damping is necessary to prevent oscillation because less deviation can occur in a shorter time, requiring less corrective input and less overshoot before correction the other way is triggered. A significant consideration when designing a null seeking trial is to use outcome measures that are as short term as practicable.

There are other clinical situations where the delay in getting results is much shorter than for brain tumours. One such situation is normal pressure hydrocephalus. In this condition, the water spaces of the brain are enlarged with the loss of brain tissue. It causes dementia and impairment of walking. In some cases these symptoms improve after a surgical operation to divert cerebrospinal fluid (CSF) away from the enlarged spaces. When the operation is successful it is usually clear within a few days, or at most weeks, that the symptoms have improved. The likelihood and degree of improvement is thought to be related to the degree to which the CSF spaces are expanded. If large, improvement is likely. If modest, it is more likely that the dementia is being caused by another condition and improvement is less likely. It would be feasible to design a null seeking trial where the index of indication included the volume of the CSF spaces as seen on a brain scan. People in equipoise would be randomized to receive a CSF diverting shunt or not, and the result would become available shortly afterwards.

Damping

The null seeking trial has a feature that most negative feedback systems do not have –memory. When driving a car down the road, oscillations from the middle of the lane occur throughout the journey. The null seeking trial does not behave in this way as it has an inbuilt mechanism to effectively damp it. With this mechanism, as time goes by data that inform the location of equipoise are collected, but never lost. As the number of cases in the trial increases, so does the precision with which the null point is defined.

The first 100 cases recruited into the trial will have a substantial impact on opinion, as no other data are available. When 900 cases have been recruited, the difference made to opinion by a further 100 cases will be smaller.

In a situation with high gain and a long time delay, a single oscillation could occur. If we were studying the effects of surgery on survival with low-grade brain tumours, it may take 5 years before sufficient results are available to influence opinion.

Substantial numbers of cases could be recruited in this time. If the negative feedback loop has high gain, and early results show that surgery is considerably more beneficial than expected, equipoise would shift towards a more aggressive application of surgery. There would be little data to prevent it from overshooting the null point. It would take further years before this became apparent, in which time many patients would be recruited. What would not happen is another overshoot away from surgery because the data collected in the first part of a trial would still be present. It is desirable to reduce this overshoot tendency as much as possible and so minimize the time delay to results becoming available. I discuss means of doing this in Chapter 9.

Correspondence of null seeking with binary trials

For a new theory to be credible it must adequately explain existing observations. If it seeks to supplement or improve on existing theory that is well supported in evidence it must reduce to the existing theory under appropriate constraints. This property of new theories is called correspondence with existing theory. A model is a theory and the model used by the null seeking trial is not new, but its use in a randomised research method is. It will not be credible if it does not correspond with the binary model that has been so successful. It does correspond. Under the constraint that the time delay before results become available is longer than the duration of the trial, negative feedback does not act at all and the scheme reduces to a binary trial. This means that binary trials are a special case of null seeking trials. Fundamental to understanding null seeking trials is that they are *not* a special case of binary trials. The null seeking trial model and method are more general that the binary, including it but not being included by it.

Chapter 8

Mathematical Model

At its simplest, the null seeking trial aims to find the crossing point of two lines. In practice the positions of the lines close to the crossing point is also important because the true null point is not likely to be the same as the crossing point but close to it, where there is a small benefit for which ever treatment is more onerous to undergo.

The mathematical model used starts with finding these lines from the data points. There are two different approaches I will present. The first of these is parametric, it is mainly intended to illustrate the method and was used in the simulation described in Chapter 6. The second is non-parametric and is likely to be the preferred method in most cases, but the complexity of non-parametric regression tends to obscure the basic method (this is discussed in Chapter 9).

For both parametric and non-parametric methods the process adopted will take the following steps:

1) Find two functions, that give estimated outcome as a function of the index of indication, that are the best fit to the data points for the two treatments being compared. This is a potentially complex task because neither the form of the function (straight line, curve etc.) or its parameters (steepness, zero crossing etc.) are known beforehand.

2) Estimate the probability distributions of the parameters of these two functions. In the simple example of straight lines this would involve finding the probability distributions of the slope and axis crossing point of the lines.

3) For all points in the x–y plane, use the probability distributions from above to estimate the probability that the treatment at point (x,y) lies above both functions (I call this probability $P1$) as well as the probability that it lies below both ($P2$). This immediately gives the probability
$Q=1-(P1+P2)$ that it lies between them.

4) When the probability that the point lies between the functions has been defined for all points of the plane, contour lines can be drawn to connect points of equal probability. The 0.9, 0.95 and 0.99 would be particularly significant contours representing the 90, 95 and 99% confidence limits that functions are different at that value of x.

In practice the steps above are intractable analytically for most treatment functions. The figures in Chapter 6 used these principles but employed simulations and "boot-strapping" (see Chapter 11) to derive the graphs shown.

The two treatments being compared are referred to as S and M, as before. For exactness we define the functions that the lines for treatments S and M represent separately from the treatments themselves. For this reason I will assign the symbol Ψ to the function that relates the outcome to the index x for treatment S, and the symbol Φ for treatment M.

Functions Ψ and Φ

The purpose of the trial is to find the crossing point of the two functions which I call Ψ and Φ. These are both functions of a single numerical parameter x, which may be a simple variable such as age or a function of several variables. In the general case:

$x = f(a_1, a_2, a_3 \ldots.. a_n)$

where $a_1, a_2, a_3 \ldots.. a_n$ are the variables, such as age or Karnofsky performance score etc. that are collected on each patient.

Further analysis is sensitive to the form of the functions Ψ and Φ. At the beginning of the trial they are assumed be linear. A system of compoundable functions such as a polynomial series could be used. Ψ and Φ will be complicated (by adding higher order terms) according to specific criteria that are unlikely to be met before the trial is well advanced. These are discussed below.

Confidence limits

The system used is to calculate the probability that points in the *x*-*y* plane lie between Ψ and Φ. Values of *x* where there are no points with a high probability of lying between $\Psi(x)$ and $\Phi(x)$ can be interpreted as within the confidence limits and values of *x* where such points do exist are without the limits. This gives a confidence range of *x*, but further information is available. The same calculation can be used as a guide to the expected difference between $\Psi(x)$ and $\Phi(x)$ for all values of *x*. What is plotted is not just the range of x where there is no significant difference but also contours of the probabilities that points across the *x*–*y* plane lie between $\Psi(x)$ and $\Phi(x)$.

Function Q

For the purpose of calculating these contours, a function is defined as Q where $Q(x, y)$ is the probability that a point with coordinates (x,y) lies either above Φ and below Ψ, or above Ψ and below Φ, but not above both or below both. In other words it is in between them. This is the probability that y is between $\Psi(x)$ and $\Phi(x)$:

$$Q(x,y)=P(y<\Psi(x))P(y>\Phi(x))+P(y>\Psi(x))P(y<\phi(x))$$
$$Q(x,-\infty)=0$$
$$Q(x,+\infty)=0$$
$$Q\rightarrow 1\ when\ P(y<\Psi(x))\ P(y<\Phi(x))+P(y>\Psi(x))P(y>\phi(x))\rightarrow 0$$
$$If\ \Psi=\Phi\ then Q\rightarrow 1\ when P(y<\Phi(x))+P(y>\Psi(x))\rightarrow 0$$

which is not possible, so:

$$Q(x,y)\rightarrow 1\Rightarrow P(\Psi(x)=\Phi(x))\rightarrow 0$$

This means that where Q takes values close to 1 there is a low probability that $\Psi(x)$ and $\Phi(x)$ are the same. Where such an area exists it will help interpretation of the evolving results of the trial and inform the position of the zone of equipoise.

Q has the following properties:

a) At any x value where points exist with Q close to 1, Ψ and Φ can be said to be significantly different.

b) Any point with Q close to 0 is likely to be above both Ψ and Φ or below both Ψ and Φ.

c) At the crossing point of Ψ and Φ, Q=½ (see next section).

d) Where Ψ and Φ are likely to be separate, the value of $Q(x,y)$ is dominated by the proximity of (x,y) to the nearest of Ψ and Φ. Probabilities pertaining to the furthest of Ψ and Φ will either be very close to 1 or very close to 0. This means that Q will give approximate confidence limits of the nearest of Ψ and Φ in such situations.

Transition function T

The purpose of the null seeking trial is to find the point of no difference between treatments. It is not primarily to define the areas in which treatments are significantly different, but it is necessary to estimate the functions Ψ and Φ. In order to do this, use can be made of the non-randomized results. In order to assess the reliability of this use of non-randomized results, transition functions T can be defined. These will show if any allocation bias is occurring that is not modelled by the functions Ψ and Φ. The transition functions will compare non-randomized data with randomized data within the two treatment groups. The mathematical approach is the same as for the function Q. For this purpose the functions Ψ_R, Ψ_S, Φ_R and Φ_S can be defined as:

$\Psi_R(x) = \Psi(x)$ on the subgroup of cases randomized to have treatment S
$\Psi_S(x) = \Psi(x)$ on the subgroup of cases selected to have treatment S
$\Phi_R(x) = \Phi(x)$ the subgroup of cases randomized to have treatment M
$\Phi_S(x) = \Phi(x)$ on the subgroup of cases selected to have treatment M

The transition functions T_Ψ and T_Φ then are:

$$T_\Psi(x,y)=P(y<\Psi_S(x))P(y>\Psi_R(x))+P(y>\Psi_S(x))P(y<\Psi_R(x))$$
$$=1-P(y<\Psi_S(x))P(y<\Psi_R(x))-P(y>\Psi_S(x))P(y>\Psi_R(x))$$
$$T_\Phi(x,y)=P(y<\Phi_S(x))P(y>\Phi_R(x))+P(y>\Phi_S(x))P(y<\Phi_R(x))$$
$$=1-P(y<\Phi_S(x))P(y<\Phi_R(x))-P(y>\Phi_S(x))P(y>\Phi_R(x))$$

These functions take values >½ if a selection effect is operating:

If $\Psi_S \equiv \Psi_R$
Then $T_\Psi = 1-P(y<\Psi_R(x))-P(y>\Psi_R(x))$
$P(y>\Psi_R(x))=1-P(y<\Psi_R(x))$
Let $P(y>\Psi(x))=z$

$$\therefore\ T_\Psi = 2z(1-z)$$

$$\therefore\ T_\Psi' = 2-4z$$

For maximum $z = 1/2$ & $T_{\Psi}'' = -4$

$$\therefore \quad T_{\Psi\,Max} = 1/2$$

This means that if there is no selection effect and $\Psi_R = \Psi_S$ then T_Ψ takes values between 0 and ½]. If T_Ψ has a value over ½, selection may be occurring. If a large selection effect is occurring T_Ψ will approach 1. At what level T_Ψ should be considered significant is uncertain. 0.95 is conventionally used for significance testing to avoid the error of deciding against a true null hypothesis. The error to avoid in this context is deciding in favour of a false null hypothesis. The figure of 0.8 is often used for this purpose. An illustrative example is shown in Figure 8.1.

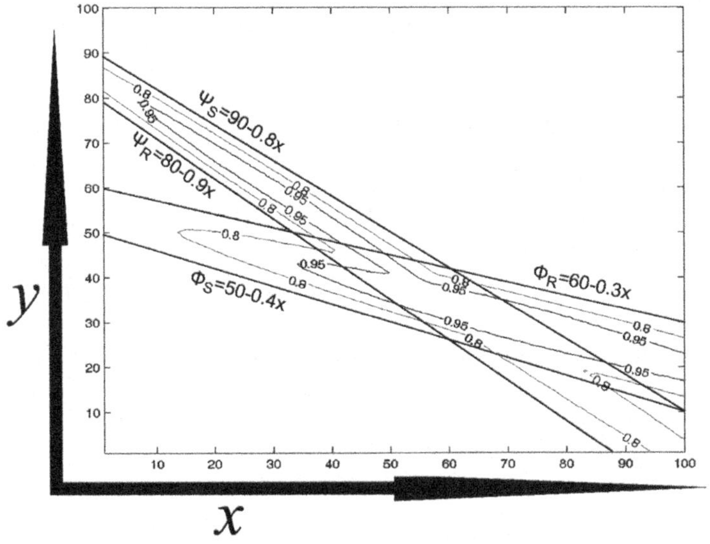

Figure 8.1 Contour plot of the transition function T for the functions shown. Note the appearance of high probability areas (0.8 and 0.95 contours) indicating that selection may be occurring regionally.

Order of Ψ and Φ

If parametric regression is used, a null seeking trial could start with the assumption that both Ψ and Φ are first-order linear functions of the index of indication x. With most regression analysis this assumption suffices. Within the null seeking trial this would also hold true when the number of recruited patients is small. The relationship between the complication rates of surgery and parameters such as patient age are known to be non-linear, as exemplified by craniotomy and clipping of cerebral aneurysms[24]. As the surgical complication rate is a significant contributor to the form of the function Ψ, it must be anticipated that there may be non-linearities in this relationship. It would not be possible reliably to model the relationship in a non-linear fashion when the number of recruited patients is small, but as it rises it may become so. Raising the order of the functions Ψ and Φ would depend on the relationship between the sum of squared residuals and the order of the function. With any data set, when the order of a regression function is raised the sum of squared residuals will fall, but a threshold level could be set at a point where the p value of the statement *the relationship is of order n rather than n+1* falls below perhaps 0.05. If the squared residuals fall below this threshold, the order could be raised by 1. The test could then be repeated until increasing the order of the equation no longer causes the sum of residuals to fall below the threshold.

Chapter 9

Non-parametric Method

Parametric regression analysis makes an assumption about the type of function that relates the variable and covariates of interest and about the type of distribution the residuals have around this function. There are numerous types of function to choose from including linear, quadratic, exponential and hyperbolic functions. They all have a general type and specific *parameters*. A straight-line relationship is of the type $y=a(x) + b$ where a and b are the parameters. Conventional parametric regression examines the data set to find the parameters, but it gives us no help with selecting the function type, which we have to assume – and we usually assume it to be linear. This is a strong and often poorly justified assumption. Numerous relationships that could form the basis of null seeking trials can be found in the literature, such as those between income and academic achievement, unemployment and crime levels or quantifiable risk factors and disease severity. Such relationships are rarely linear.

The linear regression method described in Chapter 8 can be extended to accommodate non-linear relationships, yet this is still not entirely satisfactory because it still requires that we assume the correct type of function to fit the data.

Situations in which it is reasonable to assume the type of a function for parametric regression include those where there is a good theoretical reason to believe a relationship is of that type. For example, we would expect the relationship between personal income and expenditure to be linear. Such theoretical relationships are rare in the kinds of problems that the null seeking trial is aiming to address. Another group of cases where the type of function can reasonably be assumed is when observations are distributed in a narrow band around a recognizable shape of curve, but this is also rare in relation to null seeking trials.

Non-parametric regression methods have the advantage of making no assumptions about the type of relationship between variables being studied. This in general makes them the first method to try in null seeking trials. Simple non-parametric regression is also known as scatter-plot smoothing and functions to do this are available via various statistical and spreadsheet software packages.

Other assumptions of parametric regression concern the distribution of residuals around the regression function. They are assumed to follow one of the theoretical distribution types, usually the *Gaussian* or *normal* distribution. It is also assumed that they have constant variance from one end of the regression function to the other, and that the observations are independent of each other.

Like the assumptions about the type of function, assumptions about the type of distribution are also parametric because we assume a distribution (usually normal) and we then examine the data to assess the *parameters* of that distribution, namely its variance and mean. We then make calculations from the theoretical normal distribution with our calculated mean and variance to estimate things like statistical significance and confidence limits. With purely non-parametric methods we instead calculate significance and confidence limits directly from the data, rather than via the intermediary of an assumed theoretical distribution.

Assumptions about the type of *distribution* are far less problematic than assumptions about the type of *function* mentioned above. This is largely down to the appearance of the normal distribution in the Central Limit Theorem[3]. As a consequence, variables that have multi-factorial and ill-defined determinants will

---

3 The central limit theorem states that the distribution of the sum of *n* independent and identically distributed random variables tends to a Normal distribution whatever the original distribution was.

commonly be found to have distributions that are either normal, close to normal, or easily transformed[4] to normal. Assuming a normal distribution is therefore often reasonable and has the advantage of considerably increasing statistical power, especially with relatively small data sets. Perhaps the title of this chapter should be "Partly Parametric Method" as I will later make these parametric assumptions about distribution in the course of calculating confidence and significance measures, but it is possible to avoid even these.

Non-parametric regression, or *smoothing,* is not as widely known as parametric regression and fewer options for it are available in common software packages so I shall give more background detail than with simple regression.

One of the simplest forms of non-parametric regression is the moving average estimator. This can be improved considerably, but the simple moving average enables us to study many of the decisions necessary for optimizing more complex non-parametric regression systems.

Consider a data set with n data points (x_i, y_i) for $i = 1$ to n. We can fit a smoothing curve to this data set by taking a series of x values which we label x_r across the range of x_i.. For each value of x_r we calculate the average of the y_is of data points with x_i close to x_r The s of y_is here means the pleural of y_i — not another variable. I use the same convention repeatedly. We can either average those points with x_i within a chosen *range* of x_r or we can average a chosen number of *nearest neighbours* of x_r. If we average points

4 It frequently occurs that a distribution that would otherwise be Normal is distorted by an overriding factor. One common factor is that the data are bounded, often by 0. The Normal distribution is symmetric. A variable such as peoples' body weight extends from several times the average on the high side, but it cannot be lower than 0 on the low side. It is not distributed symmetrically and so not Normally. A simple mathematical function such as the square root of weight is closer to a Normal distribution than weight. This is a transformation and it allows statistics that assume a Normal distribution to be applied to the transformed data more rigorously.

within a range of x_r the range is called the *bandwidth* of the moving average estimator.

If we use the nearest neighbours option, the number of nearest neighbours averaged is defined by the *span* of the moving average estimator. By convention the span refers not to the actual number of points included but to the proportion of total points included.

Span values are therefore between 0 and 1. We may either use values of x_r regularly spaced across the range of x_is or use x_r s that are the same as the x_i s.

Figure 9.1 shows a scatter plot of 100 data points that were created by adding a uniformly distributed random variable to an underlying function. In real situations the underlying function is what we are trying to estimate and is not known at the beginning, but for illustration it helps us tell how good the non-parametric system used is at recovering the original function.

Figure 9.1a shows the scatter plot with the underlying function which is part of the curve of:

$$y = x^2 \sin(x)$$

In the other figures the true function is shown as a fine line. Figures 9.1b–d show the same data with moving average non-parametric regression lines added. The difference between the graphs is the span of the moving average. Figure 9.1b shows a moving average with a span of 0.09. The y value of each point on the regression line is equal to the average of the y values of the 9 data points with x values nearest to that value of x_r. The spans of Figures 9.1c and 9.1d are 0.23 and 0.4 respectively. They illustrate that the smaller the span the less smooth the curve is, but the random variations that do occur faithfully centre around the underlying function. The larger the span the smoother the curve, but regions emerge where it systematically deviates from the underlying function. As the underlying function is not known in real situations, increasing the span can give the misleading impression of a credible function, which is actually untrue. The random unevenness of the curve seen with small spans is known as *variance* and the systematic deviation between the curve and the real underlying function seen with large spans is known as *bias*.

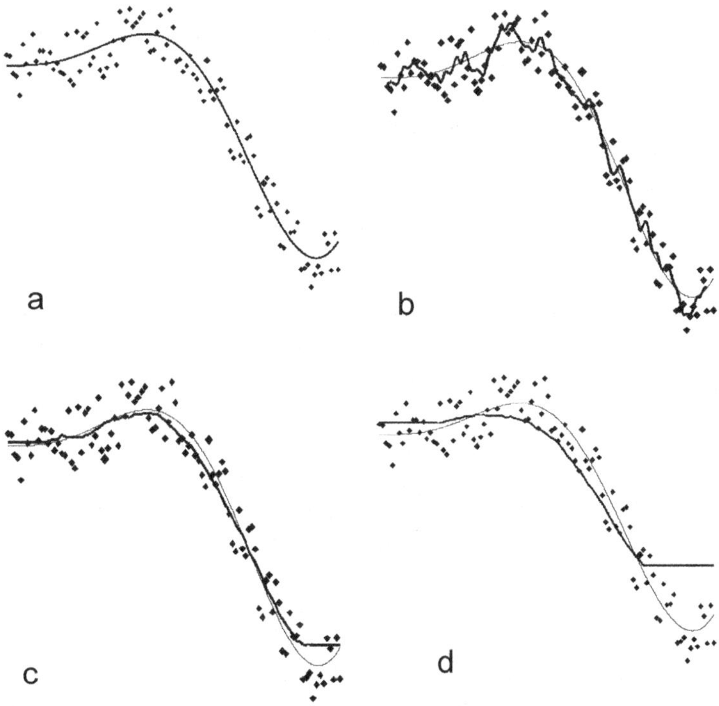

Figure 9.1. Simple nearest neighbour moving average estimators. (a) The original function $y = x^2 \sin(x)$. The 100 data points are simulated by adding a random variable. (b) The original function (fine line) and the moving average estimator using a span of 0.09. This is increased to 0.23 in (c) and 0.4 in (d). Note that the estimator becomes smoother with increased span, but bias also increases. The boundary bias on the right-hand side in (d) is particularly severe.

The simple moving average estimator can be improved in several ways but these basic relationships between span or bandwidth, variance and bias recur with all forms of non-parametric regression. Another limitation of moving average estimators is that they suffer from boundary effects. The problem arises because in the middle of the data set an average can be taken of several data points on either side of x_r but at the boundaries on the left- and right-hand ends of the graph they can't because there are no data points on the left-hand side of the left-most point or on the right-hand side of the right-most point. This problem can be dealt with in several ways.

We could simply not display the moving average in areas where it is not symmetrical about x_r but this means leaving no regression line near the boundaries. We could display the moving average near the boundaries, but reduce the span so that it remains symmetric about x_r yet averages fewer values – this is the so-called *symmetric neighbourhood* method but it increases the variance of the estimate near the boundaries. A commonly used method is to average the nearest neighbours whatever side they are on. This means that the average for the left-most side of the graph is of points that are on the right of x_r and for the right-most side all the nearest neighbours are on the left. In the middle of the graph they are divided between the two sides. Also the nearest neighbours for the first few data points are the same, so the curve is flat as is seen in Figure 9.1. This flattening at the boundaries is called *boundary bias* and, like bias, it is in general worse for larger spans.

The first improvement we can make to the simple moving average estimator is to weight the averages according to how far data points are from x_r. This improves the compromise between variance and bias by allowing the span of the moving average to be increased while still reducing the amount of bias because points far from x_r have less influence on the average. This method of weighted averages is known as *kernel estimation*. Several ways of weighting the data points in the average can be used including the tri-cubic and normal functions. A simple moving average has a rectangular function. These three commonly used kernel functions are shown in Figure 9.2.

Kernel functions have defined bandwidths but not defined spans; that is, they include data points with x_is within a given range of x_r irrespective of how many such points there are. This can be problematic because variance will increase in regions with few data points and if there are no data points within the bandwidth we will have no estimate at all. To get round this problem we need a convenient way of altering the bandwidth of the kernel estimator to ensure a constant or at least a minimum span. To do this we put a multiplier h in the distance measure $(x_i\text{-}x_r)h$ before applying the kernel weighting function. If we find too few data points within the bandwidth around x_r we reduce h so more distant points come within the bandwidth. Conversely if we are using a constant span method and find too many data points within the bandwidth of x_r we increase h.

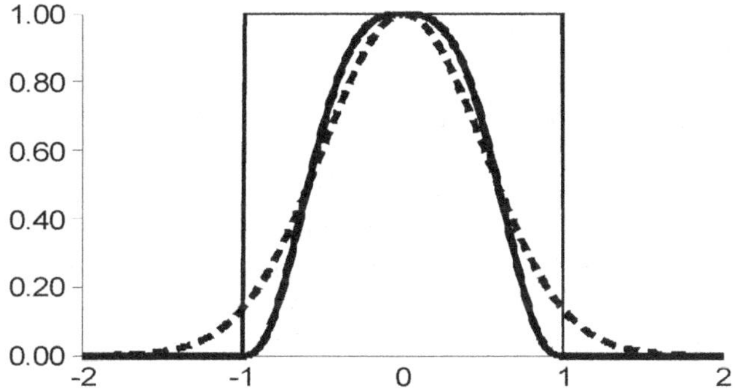

Figure 9.2
Tricube (heavy black curve), normal (dotted curve) and step (fine black line) kernel functions. The standard normal function $\frac{1}{\sqrt{2\pi}}e^{-x^2/2}$ *has been scaled by a factor of ½ in the x direction and* $\sqrt{2\pi}$ *in the y direction to become* e^{-2x^2}.

Kernel estimates improve smoothness while reducing bias and they lessen boundary bias because although all the points being

averaged are the same for x_i near the boundaries, their weighting changes as the boundary is approached. The kernel estimator K is a function that returns a weight w for individual data points, depending on their distance in the x direction from x_r.
If the data points are (x_i,y_i) for $i = [1:n]$
Then:

$$w_i = K[(x_i - x_r)h]$$

The usual normal kernel function is:

$$K_N(x_i - x_r) = e^{-2(x_i - x_r)^2}$$

The tricube kernel function is:

$$for\ |z|<1, K_T(x_j - x_r) = (1-|(x_j - x_r)|^3)^3$$
$$for\ |z|\geq 1, K_T(x_j - x_r) = 0$$

The average is a simple summary statistic that can be weighted using a kernel function, but it is not the only one possible. Many other statistics can be also weighted using the kernel. The ones that are most relevant to our present purpose are local regression functions, discussed next, and Kaplan-Meier-type survival functions.

Staying with the weighted average for now, the estimated y value of the regression function at x_r is then:

$$\hat{y}_r = \frac{\sum_{j=1}^{n} w_j y_j}{\sum_{j=1}^{n} w_j}$$

The circumflex on the $\hat{y}_r$ is used to show that it is an estimate rather than a value from a data point in the data set y_i, or the y value of the true underlying function y_r.

Local linear regression

The next improvement we can make is to replace the weighted average of the data points within a bandwidth of x_r with a least-squares function-fitting process with the same weighted data points. Least-squares regression can be applied to any type of function. Most commonly it is applied to straight-linc or linear functions, when it is called linear regression. Linear functions are the special first-order case of the more general class of polynomial functions and local regression is often referred to as local polynomial regression to keep the argument general. Polynomial functions have the form:

$$y = a_0 + a_1 x + a_2 x^2 + \ldots + a_n x^n + \varepsilon$$

Where n is the *degree* of the polynomial. If n=0 the fit is to a single numerical value (the average value of the data points). If n=1 it is linear regression. If n=2 it is quadratic regression, and so on. ε is the error term that refers to the *residual* difference between the data points' y values y_i and $\hat{y}_r$. In practice the terms "linear" and "polynomial" are largely synonymous as higher-order local polynomial regression is rarely justified. Least-squares regression can be done using kernel weighting of the data points according to their distance from x_r when it is called local regression.

To fit a linear function at a local point x_r we are not interested in the actual x values of the data points but rather their x values relative to x_r so the function becomes:

$$y = a_0 + a_1 (x_i - x_r) + \varepsilon_i$$

To fit a local linear regression line to the data around x_r requires several computational steps. First is to apply the kernel function.

$$w_i = K[(x_i - x_r)h]$$

to all points in the data set. This will give a value of or close to 0 if point x_i is outside the kernel bandwidth around x_r and a weighting value (w_i) between 0 and 1 if it is within the bandwidth.

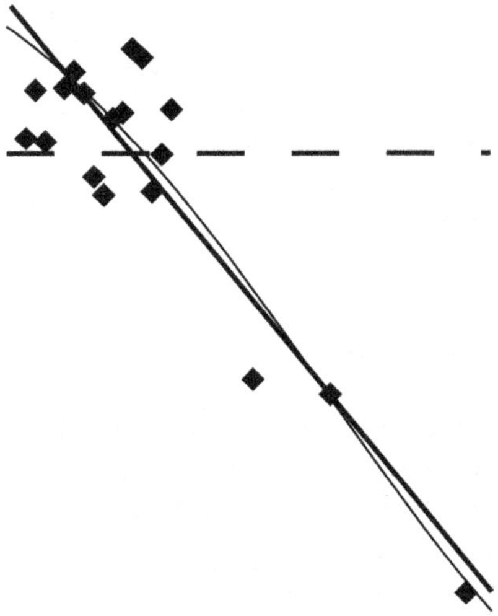

Figure 9.3. The advantage of local linear regression over local averaging. It is most pronounced when the data points are not evenly distributed in the *x* (horizontal) direction across the kernel bandwidth. In this case they are concentrated at the left-hand side. The fine line is the original function. The heavy black line is the local linear regression. The dashed line is the average of the *y* values. If we are calculating the non-parametric regression function in the middle of kernel band, the linear regression gives a less biased estimate than the average.

The process of conventional least-squares regression is then adapted to minimize the sum of weighted residuals $\sum_{i=1}^{n} w_i^2 \varepsilon_i^2$ rather than the conventional simple sum of residuals $\sum_{j=1}^{n} \varepsilon_j^2$. When the sum of weighted residual squares has been minimized to obtain the fitted line through x_r the value that is used in the estimator is only the constant a_0 as the other term is 0 where $x=x_r$. This process is repeated for all x_rs where a regression estimate is required. Joining the points gives the non-parametric regression function.

The reason why local linear regression helps is illustrated in Figure 9.3. The kernel process suffers from biasing error if it happens that observations are not evenly distributed across the x_rs in the bandwidth, most markedly so in the boundary situation. Local linear estimators are widely used to assist the interpretation of scatter plots. A key parameter that has to be chosen is the band width of the kernel. This is normally guided by the visual impression of the graph, and in selecting we often start with 0.5 and work towards the smallest that gives an acceptably smooth result.

This approach requires direct input from a statistician, which makes it unsuitable for null seeking trials where we need the regressions to be done automatically by computer for interpretation by people who are not necessarily skilled in the methods of scatter plot smoothing. We therefore need to take the subject further to automatic methods of selecting the span of the estimator. Recall that the span of an estimator is the proportion of data points that contribute to the estimate. The bandwidth h is the distance from x_r over which points are included.

Several methods have been described for optimizing the bandwidth of local polynomial regression models. We will stick to a local linear model, although the results can be generalized. These methods all use properties of the estimated regression function itself in order to estimate optimal bandwidth, so it is necessary to start with a first-approximation regression function, then use it to derive such things as its rate of change of curvature (second-order derivative or $\hat{y}_r''$) and the residual variance of the data around it.

These values are then used to improve the selection of bandwidth and hence improve the regression fit. In general several iterations are necessary, starting with an intermediate value constant bandwidth such as 0.5, and using this to improve the regression to a second approximation. Then this second approximation is used to recalculate the above derivatives in order to produce a third approximation, and so on, until two successive approximations are acceptably close together. Furthermore the processes outlined below are done at each point where the regression curve is to be estimated, because optimization may involve replacing a constant bandwidth with one that is changed for different values of x_r. This all adds up to a large amount of iterative calculation, but with modern computers it is tractable.

The first method of span optimization depends on standard formulae[27]:

The expected value at x_r is given by:

$$E(\hat{y}|x_r) \simeq \hat{y}(x_r) + \frac{h^2}{2} s_K^2 \hat{y}''(x_r)$$

he variance at x_r is given by:

$$Var(\hat{y}|x_r) \approx \frac{\sigma^2 a_K^2}{nh} \hat{y}(x_r)$$

The bias at x_r is given by:

$$Bias(\hat{y}|x_r) \approx E(\hat{y}|x_r) - \hat{y}(x_r) \simeq \frac{h^2}{2} s_K^2 \hat{y}''(x_r)$$

The objective is to minimize the mean square error (MSE) which is the sum of squared bias and variance. The bandwidth at x_r that minimizes the MSE is:

$$h^0(x_r) \approx \left[\frac{a_K^2 \sigma^2}{s_K^4 n f(x_r)[\hat{y}''(x_r)]^2}\right]^{1/5}$$

where

$(\hat{y}|x_r) = \hat{y}(x_r)$ is the fitted value at $x=x_r$

$\sigma^2 = V(\varepsilon)$ is the residual variance of all data around the regression function.

$h0$ is the optimal bandwidth

n is the sample size

$f(x_r)$ is the probability density function for x at x_r. If this is low, there are few data points with x values near x_r, and if it is high there are many.

sK2 and aK2 are constants derived from integration of the kernel function:
$s_K^2 = \int z^2 K(z)\, dz$ and $a_K^2 = \int [K(z)]^2 dz$

For kernel functions that are symmetric distributions around 0, s_K^2 is the variance of that distribution.

An alternative to using these formulae is to calculate the squared residuals over the whole regression plot and select a span that minimizes their average. As stated this won't work because if the span covers only 1 data point, residuals will be 0 and shorter spans will give lower residuals with no concession to smoothness.

Instead we can select the x_r s to be the same as the x_is in the data set, then for each one we can calculate the residuals for all data points omitting the single y_i value. The resulting residuals are a reasonable guide to optimizing bandwidth. This method is known as *cross validation*. In principle it can be done for various bandwidths at each point, however for large data sets the need for iterative arithmetic is substantial. The basic process is:

1) Decide on a range of bandwidths to be investigated. To be comprehensive this should be from close to 0 through to 1. For a large data set it is usually economic to investigate a small number of fairly widely spaced bandwidths as a first approximation, and then identify the region of minimum average squared residuals for investigation with higher resolution.

2) Decide on a range of x_r values at which to plot the local regressions.

3) For each value x_r where we wish to plot the regression function we omit its y_r from the data set before calculating the value of the regression function $\hat{y}_r$ for the selected bandwidths. For each bandwidth we then note the difference between the y_r we omitted and the regression value $\hat{y}_r$. This difference is then squared and the squared differences for all x_i averaged for bandwidth h at all x_i. These are then plotted and the minimum taken as a guide to the span to use for the final calculation of the regression plot.

Outliers

The method of least-squares regression is particularly sensitive to outliers as a consequence of minimizing the *square* of the distance between each point and the regression line. Local linear regression depends on using least-squares regression and so it is also sensitive to outliers. This is undesirable as outlying observations are often of questionable validity. The problem can be addressed by weighting data points according to the magnitude of their residuals. As we have seen above with optimizing the span of kernel functions, this method also requires us to make a first approximation to the regression function, in order to calculate squared residuals, then to weight the residuals and recalculate the regression curve. We continue iterating these steps until consecutive iterations produce results that are not significantly different from each other.

The weighting system used can be similar to the kernel functions. The Tukey bisquare weighting function is a popular one. It takes the value $(1-\left(\frac{e}{k}\right)^2)^2$ for $|e|\leqslant k$ and 0 for $|e|>k$. Another popular one is the Huber weighting function that takes the value 1 for $|e|\leqslant k$ and $\left(\frac{k}{|e|}\right)$ for $|e|>k$ where e is the residual and k is a tuning constant often set as 1.345σ for the Huber function and 4.6853σ for the Tukey bi-square function.

Weighting systems intended to minimize the influence of outlying data points are known as *robustness* systems. They are not always necessary. Residual plots can be inspected to see if a data set contains significant outliers. If it does not, and the data seem to be reasonably normally distributed, then robustness systems are not necessary. They also carry a small cost in accuracy. For the purposes of null seeking trials it is important to make the analysis system automatic. Inspecting residual plots is a manual process that is difficult to automate, for which reason it is suggested that robustness weighting be included.

This system of non-parametric regression using local linear regression, tri-cubic kernel weights and bi-square robustness weights was introduced by Cleveland in 1979[28]. He called the algorithm *lowess* for *lo*cally *we*ighted *s*catter plot *s*moothing. He later developed a generalized method for multiple regression and renamed it *loess* for *lo*cal regr*ess*ion[29]. The relevance of these terms is that there are several statistical software packages (such as the open-source package R®) which offer these functions under these command names.

The first step in analyzing the results of a null seeking trial is to calculate reasonable non-parametric plots to fit the data from the two groups with treatments M and S. This will be a guide to where the currently estimated null point is, but it gives no indication about confidence in the accuracy of the plot or whether the difference between the two plots is significant or not in different areas. For this additional information one option is to plot confidence limits around the regression curves. Another way is to plot a second function of significance alongside the two non-parametric regression plots.

Parametric v non-parametric methods for null seeking trials

Parametric regression is a well-developed and competent subject and there are numerous methods available for curve-fitting by optimizing the choice of function and transforming the data to make it fit a simple relationship or standard distribution. Powerful though these tools are, they introduce a further specific problem for our present purpose. Their use, and the interpretation of the results they give, often requires considerable statistical expertise. A null seeking trial works by creating a negative feedback loop, an essential component of which is the presentation of data to, and its interpretation by, individuals who may have little knowledge of complex statistics. In this situation a method that simply tries to estimate the most representative and reasonable-looking curve through a scatter plot has the advantage of being easily accessible and comparatively transparent. It is for this reason that I would tend to prefer non-parametric regression, at least in the first instance, over elaborate parametric non-linear regression techniques.

Parametric methods have advantages that make them the methods of choice in cases where the necessary assumptions can be justified. A fairly rigorous logical argument, free from arbitrary decisions, leads from the assumptions to not only the regression function parameters, but also to other useful derivatives such as confidence limits and significance levels. In contrast to this, non-parametric methods avoid the problematic assumption of parametric methods, but the argument connecting the observed data with the regression function and associated statistics is not rigorous, depending as it does on many arbitrary decisions, such as what the span should be and how smooth a curve is smooth enough. Then there is the problem of bias which confounds extrapolating results from the highly idealized case where the true underlying function of the data set is known.

Parametric methods generally give narrower confidence limits and greater power than non-parametric methods used to analyze the same data set. This advantage of parametric methods is less important because confidence and power calculations done on non-parametric regression functions are not usually non-parametric. Instead they use parametric assumptions of distributions, as these are more easily justified than assumptions about the form of the regression function itself.

Chapter 10

Confidence and Significance

The non-parametric regression methods described in Chapter 9 are used in null seeking trials to give estimates of the relationship between outcome and indication variables. They do not assess our confidence in the position of these estimates or the range over which the true relationships may lie while still being consistent with the data observed. These issues are addressed by calculations of confidence limits and tests of statistical significance.

As a first approximation, the purpose of a null seeking trial is to determine the point at which two treatments are equivalent and, as such, a natural approach to analysis is to calculate a confidence range for the position of the crossing point of the two treatment regression functions. Simple as this is, it cannot be assumed in the general case. To illustrate this we return to the surgical example given in Chapter 6. In this situation a surgical operation is being compared with no operation. The operation is quite a major one, involving substantial cost, discomfort, risk and inconvenience. If we could confidently say to a patient that the outcome from the operation was indistinguishable from the outcome following no operation, equipoise would not exist. We would weigh the cost, the risk and the discomfort of the operation on the one hand, against no benefit on the other, and unanimously conclude not to operate. There would be no equipoise and no randomization. The region of equipoise would lie to the left of the crossing point and the judgement regarding equivalence would be to consider how much benefit from the operation balanced its costs. In order to make this judgement, a quantitative estimate of the operation's benefit would be highly relevant.

It would then be more important to estimate the regression function and its confidence limits, than to estimate confidence limits around the zero crossing point. Most relevant, would be to estimate confidence limits around the difference between the two treatment functions.

In practice, both methods of analysis will have their applications and a generally applicable scheme of analysis can include both. As with much of the analysis of null seeking trials, the estimations are complex, involving compound probability distributions. Estimating 95% confidence limits around a simple non-parametric regression curve involves well-established techniques. What we require is confidence limits on the difference between two curves rather than specifically around each one.

The difference between two curves

The basic method of calculating these confidence limits is to start with estimates of the probability distributions of the positions of the two curves at a series of points of interest. We then take a compound distribution of the difference between them. In principle, this means taking each possible position of one curve with its assigned probability and calculating the combined probability of that with all the possible positions of the other curve. This process is repeated at all possible positions of the first curve.

Fully To calculate such compound distributions is demanding in terms of computer time, so short-cuts are taken to make the problem more tractable. The first short-cut is to truncate the probability distributions of the positions of the two curves so that we do not bother calculating the combinations of probabilities where one of them is extremely remote. When we have finished our calculations, we check how much accuracy is sacrificed by this truncation by adding up the probabilities of all the combinations considered. A complete analysis would give a sum of probabilities of 1. Truncation in this way reduces the total sum below 1 but this loss of probability is tolerable if the reduction below 1 is trivial. Just how trivial is a matter of judgement. If the sum is more than

0.999 it is likely that any confidence limits calculated in the more familiar 0.95 region will be accurate enough for practical purposes.

The first option of plotting confidence limits is a derivative of the conventional way of drawing up confidence limits around a non-parametric regression. The significant limitation of this method is that for small data sets it relies on a parametric assumption that the distribution of data around the regression line is normal and has a constant variance for all values of x_r. This assumption is often not so unreasonable, as the principal problem with regression analysis in this context is the assumption of the form of the function – not of the distribution of data around it. These calculations on the difference between functions require probability distributions of the functions themselves. Defining these distributions is the penultimate step in the more familiar task of calculating confidence limits, or more generally *bands* around those functions, to which we now turn.

Confidence envelope.

As a step in the development of confidence bands around a non-parametric regression estimate, we make some generally fairly justifiable assumptions. First, we assume that the variance of residuals around the estimate is independent of the value of x_r. Second, it is convenient to assume that the distribution is normal.

The fitted values $\hat{y}_i$ in local polynomial regression are the sums of all weighted observed values that fall within the kernel:

$$\hat{y}_i = \sum_{j=1}^{n} w_{ij} y_i$$

This means that $\hat{y}_i$ is a linear function of the y_is. The variance of $\hat{y}_i$ can be simply calculated by replacing the values of y_i with an estimate of the variance of those values. As we assume that the variance of the whole data set is constant and not dependent on x_r

we can estimate this by assigning the overall variance at each $\hat{y}_i$ within the kernel:

$$V(\hat{y}_i)=\sum_{j=1}^{n}\sigma^2 w_{ij}=\sigma^2\sum_{j=1}^{n}w_{ij}$$

then approximate 95% confidence limits are:

$$\hat{y}_i \pm 2\sqrt{V(\hat{y}_i)}$$

All we need to apply this is a value for σ^2. Unfortunately such a value is not so easy to find. Variance is the average of squared residuals around a regression function, but in the case of non-parametric regression, variance is dependent on our choice of bandwidth and is reduced by reducing the bandwidth because then the regression estimate follows the data points more closely (although with more variability itself). A further factor that complicates the analogy with linear regression is the effect of bias, which inflates the sum of squared residuals. Another problem is the number of degrees of freedom. In the case of parametric regression analysis the method of calculating variance is straightforward. We divide the sum of squared residuals by the number of degrees of freedom of the model. The degrees of freedom are, in turn, the number of data points minus the number of parameters necessary to define the regression function. In the case of a linear function this is 2, for a quadratic it is 3 and so on. The number of degrees of freedom to use for non-parametric regression estimates is not so obvious.

One way round these problems is to use pseudo-residuals, which do not depend on the regression function but are calculated from the differences between consecutive data points. This approach was used by Rice in 1984 [30]. The estimated variance is:

$$\hat{\sigma}^2=\frac{1}{2(n-1)}\sum_{i=2}^{n}(y_i-y_{i-1})^2$$

A problem with this method is that the difference between consecutive data points depends on the trend in the underlying function. Not taking account of this trend tends to inflate the variance. This effect is most marked in steeply sloping functions with sparse data points. An alternative method, proposed by Gasser et. al. in 1986 [31] largely solved this problem. Their method is a form of interpolation with pseudo-residuals defined as:

$$\varepsilon_i = \frac{x_{i+1} - x_i}{x_{i+1} - x_{i-1}} y_{i-1} + \frac{x_i - x_{i-1}}{x_{i+1} - x_{i-1}} y_{i-1} y_{i+1} - y_i$$

If this is written as $\varepsilon_i = a_i y_{i-1} + b_i y_{i+1} - y_i$

then the estimated variance is:

$$\hat{\sigma}^2 = \frac{1}{n-2} \sum_{i=1}^{n-1} c_i^2 \varepsilon_i^2$$

where $c_i^2 = (a_i^2 + b_i^2 + 1)^{-1}$

Non-parametric confidence limits

The balance between variance and bias that affects non-parametric regression systems is particularly relevant to confidence limits and their inferences. In the case of linear regression modelling, the assumption we have to make about the linear nature of the relationship between variables and covariates is a strong one, but in cases where the assumption is reasonable then other assumptions necessary to calculate meaningful significances and confidence limits fall into place fairly easily. If a linear relationship can reasonably be assumed, it is also usually reasonable to assume that the data sample is representative of the population of interest, that its variance and mean are also representative, and that the regression line we calculate is fairly close to the true relationship.

The exact statement of the probability that is calculated to obtain confidence limits varies but it is generally fair to assume that the probability that the true relationship lies outside of a 95% confidence band around the regression line is reasonably close to 0.05. Unfortunately there is no corresponding line of argument that can be applied to non-parametric regression curves.

Bias is the limiting factor in our ability to draw inferences from non-parametric regression. We can estimate the position of a regression line and then we can exactly calculate the variance of the data around that line. However, an exact knowledge of bias requires that we know what the underlying function is — which we do not. We can only estimate bias. Estimates can be quite good and we can use sophisticated methods for minimizing bias, but there are particular problems that we cannot eliminate, such as substantial bias that arises when the underlying function has a sharp corner that we don't know about, especially if it is in an area where there are few data.

The moving kernel function is a powerful tool. It is fairly straightforward to calculate means, medians, survival centiles and probability distributions with corresponding confidence limits and p values within the kernel and plot them alongside the regression line. The selection of the kernel's bandwidth, though guided by various optimization methods, is ultimately somewhat arbitrary. This arbitrary selection has a profound influence on the position of the regression line and also on its 95% confidence limits.

We should not be tempted to address this problem by reverting to linear regression. Choosing an infinite kernel bandwidth for local linear regression non-parametric estimators is equivalent to assuming a linear relationship – and that is a circumstance in which bias is maximized. Our approach will be to do our best to minimize the problems of bias, to use confidence limits as a guide to the data distribution around our regression lines but to remain mindful of the limitation of the methods. In null seeking trials this limitation is addressed by gathering more data.

Bootstrapping

Boot-strapping is a method that is in principle very simple and applicable to virtually any confidence limit or distribution problem, which makes it powerful. It is a sledgehammer approach that is intensive in computing time. It produces readily available and credible results that allow a project to proceed before other methods are finalized. Furthermore it provides a useful baseline against which to check the results of other methods.

The idea is to take a data set with n data points (x_i,y_i) where i is an integer from 1 to n, and derive from it thousands of other related data samples. The critical assumption is that the process of deriving the boot-strap sample from the data set is analogous to the process of deriving the data set from the population, making the relationship between the boot-strap samples and the data set analogous to the relationship between the data set and the population. For each boot-strap sample we calculate our selected summary statistic, such as a regression estimator or Kaplan-Meier centile position. When we have done this about 10,000 times we have enough estimates to approximate a distribution, from which we can calculate confidence limits. There are a variety of methods of producing thousands of derivatives from a data set. The most commonly used is known as *sampling with replacement*. This takes its name from situations commonly modelled in elementary probability texts, such as taking coloured balls from an opaque bag. We can remove a number of balls from the bag as a sample of the bag's content; this is sampling *without* replacement. Alternatively we can remove balls one at a time, note each one's colour and then return it to the bag before removing the next one; this is sampling *with* replacement. When sampling with replacement from a set of data points (x_i,y_i) we choose at random one of the data points, (x_j,y_j), and copy it into the sample. We then choose a second data point, also at random, but we have not removed (x_j,y_j) from the data set, so there is a chance of choosing it again. We carry on doing this until we have a sample of n data points, some of which are likely to be repeats, in which case some data points in the original set will be missing.

Sampling with replacement gives a true subset of the data that can have the same number of points as the data. There are a huge number of possible combinations such that we are likely to have enough for a thorough boot-strap analysis, unless the data set is small.

A variant of this method proposed by Hardle and Bowman[35] is to use the regression estimator to calculate residuals of each data point:

$$\varepsilon_i = y_i - \hat{y}(x_i, h)$$

where $\hat{y}(x_i, h)$ is a pilot non-parametric estimator. The bandwidth of the kernel h is included because the function is dependent on it. Because of bias, these residuals will not sum to 0 so they are normalized:

$$\tilde{\varepsilon}_i = \hat{\varepsilon}_i - \frac{1}{n}\sum_j \hat{\varepsilon}_j$$

These new residuals sum to 0. The bootstrap observations are then

$$y_i^{boot} = \hat{y}(x_i, h) + \varepsilon_i^{boot}$$

where the ε^{boot} values are taken from the $\tilde{\varepsilon}$ values by random sampling with replacement.

$$\tilde{\varepsilon} = \varepsilon - \frac{1}{n}\sum_j \varepsilon_j$$

This reduces but does not eliminate bias so we still end up with a *variance band* rather than true confidence limits.

The steps in the boot-strap procedure are:

1) Select a series of x-values for which confidence limits or p values are required.

2) Run the regression estimator on the whole data set to get the estimate at each x value.
3) Select a boot-strap sampling method (e.g. sampling with replacement).
4) 4) Derive a "sample" of the data using the selected sampling method.
5) Run the regression estimator again on the sample and note the estimates for each x-value.
6) Repeat steps 4 and 5 a large number of times, typically 10,000.
7) For each x value rank the 10,000 estimates from step 5.Take the symmetrical percentile range from these ranked 10,000 estimates as the variance limits. So, for example, if 95% limits are required then take the 2.5 and 97.5 centiles or the 250th and 9750th values in the ranked estimates as the boundaries of the range.

Chapter 11

Survival Outcome Measures and Censored Data

An important group of outcome measures are those based on time. These include the length of survival of patients after diagnosis of a mortal disease, the time intervals between attacks of an episodic disorder such as epileptic fits or the time it takes from commencing an educational programme to successfully passing an exam or achieving gainful employment. Time is a powerful outcome measure for several reasons. It can also be measured easily, reliably and accurately with very high resolution. Time-based outcomes can be assumed to start at zero, which is a powerful assumption when calculating trends and significance. Time is highly linear; a specific time interval is the same whether it is referring to one patient or another or to different periods in an individual patient's survival.

The main disadvantage of survival data is that the data sets are usually partly censored. When we analyze data sets on survival, we find that in some cases survival time is known, but in others all that is known is the last time an observation was made. In the case of patient survival following diagnosis of a mortal disease, if 100 patients are examined at 5 years we will find that a number have died and exact survival times will be available for them. Others will still be alive so we cannot say that they have a survival of 5 years – only that they have a survival of at least 5 years.

The solution to this problem involves adapting the basic idea of a kernel function with a particular span or bandwidth that is swept through the data from one side to the other in the same way as a moving average or local polynomial regression function. Instead of the moving average-based function, we require a numerical derivative of the Kaplan-Meier survival function that we can plot at each point where we want to make a non-parametric estimate. Of the many numerical derivatives that are possible, the two simple alternatives are: time to a specific survival centile; and centile at a specific time. An example of time to a specific centile is median survival, which is the time it takes for the population's survival to fall to 50%. In this situation the population is data points within the kernel. An example of centile at a specific time is the commonly used outcome measure for trials in cancer, namely the 5-year survival rate.

Either of these can be plotted as a non-parametric regression smoothed using a kernel function. Which one to use depends on the situation, and it may frequently be desirable to make both available. In cases where the outcome is recorded at specific times after the treatment was administered, survival at a specific time has the advantage of being directly and immediately available. In some of the previously described trials on the effects of carotid stenosis surgery for preventing stroke (chapter 3), outcome data was recorded at yearly intervals and patients were simply asked whether they had had any strokes in the previous year. If they had done, the date on which the strokes occurred was frequently not available. Stroke-free survival at 1 year was to hand, but the time until 50% of people had suffered a stroke was not.

Situations in which outcome events are recorded as they happen are different. In many cancer trials, deaths of patients are reported to the trial centre as they happen. This means that time to a specific survival centile and survival at a specific time are both available. In either case, after a specific time interval we always know what the survival at that time is, but we do not always know the time to specific survival centiles.

The procedure used is to define a window centred on x_r and to construct a Kaplan-Meier curve of the data falling within this window. The median, or whatever centile is chosen, can then be obtained from the Kaplan-Meier curve and plotted before moving x_r on to the next value. If the underlying centile curve is assumed to be smooth, any kernel weighted approach can be used. This method was proposed by Beran[32] in 1981 and further developed by Doksum and Yandell[33] in 1982 and Gentleman and Crowley[34] in 1991. For each value x_r, kernel weights $w_{ir}(x_i - x_r; h)$ are calculated from the surrounding data points x_i. The weighted Kaplan-Meier curve then is:

$$\hat{S}(y) = \prod_{i:\, y_i < y} \left[1 - \frac{\delta_i w_{ir}(x_i - x_r; h)}{\sum_{j \in R_i} w_{jr}(x_j - x_r; h)}\right]^{\delta_i}$$

where δ_i is 0 for censored data (e.g. patient still alive) and 1 for non-censored data (e.g. patient dead) and R_i is the set of individuals at risk at time y_i. This method can be adapted for survival at a specific time by taking the δ_is at a specific time after subjects (data points) were recruited and noting the minimum value of $\hat{S}(y)$, or time to a specific survival by taking δ_is from the latest available data and finding centiles from the resulting Kaplan-Meier curves.

This equation is a step function and will not yield a continuous range of centiles without smoothing, particularly if the data are sparse. Various smoothing techniques are possible. Perhaps the simplest is to bisect either the horizontal "treads" or the vertical "risers" of the step function and plot straight lines between the resulting points. Another option is to smooth the corner points using one of the non-parametric regression methods discussed above. From these smoothed Kaplan-Meier curves, specific centiles can be found and plotted.

As with the non-parametric regression methods discussed above, here it is necessary to make a choice of optimal kernel function and bandwidth, but there is an additional choice to be made — which centile to plot. In general, the 50th centile or median will be the first choice because it is familiar to most people and readily interpreted. In a null seeking trial it is desirable to minimize the time delay between patients being recruited and the results becoming available. If a mortal disease is being studied and the censored data approach is appropriate, values for median survival become available after half the population recruited has died. If the condition being studied has a long survival, this could lead to a delay of several years by which time substantial data may have been collected. We can improve the time delay of null seeking trials by plotting a higher centile of survival. In the case of low-grade brain tumours, the time delay before a reasonable median can be plotted is around 6 years, whereas reasonable 90th centile survival plots can be made after 18 months.

An alternative approach to shortening the time delay for results in a null seeking trial is to plot survival at a specific time, rather than time to a specific survival centile. This has the advantage that we can define the delay. In a trial of brain tumours we might start with 18 month survival and add in 3 year survival later, and use median survival finally as the data become available.

Chapter 12

Null Seeking Trial Related Methods

I am aware of no equivalent of the null seeking trial in the existing literature, probably because its objective seems perverse. Features of it have been developed in other contexts, including trials with stepped-wedge designs, sequential trials, futility analysis and adaptive designs. These techniques cannot be directly applied to the null seeking trial for two reasons. First, because they generally they seek to answer the question: *Is there a difference between two populations?* rather than the null seeking trial question of: A*t what point is there no difference?* Second, because they assume that the populations from which the samples are drawn remain constant over the course of the investigation, which is not true of a null seeking trial.

Stepped-Wedge Trials

Stepped-wedge designs are used for interventions that are thought to be beneficial, leaving no equipoise. The trial design administers the intervention to all participants but randomizes the time at which it is given. The trial then compares the time period before intervention with the time period after intervention[36]. The first trial of this kind was the Gambian hepatitis intervention study in which vaccinations were administered at a random time to all participants and the incidence of hepatic carcinoma was compared before and after vaccination[37].

The ability of stepped-wedge trials to be conducted in situations in which there is no equipoise makes their design potentially useful in surgery. It would be largely restricted to conditions in which there is a long enough window of opportunity for surgical treatment to allow randomization of the intervention time. Urgent operations, or those that are aimed at short-term palliation, could not be easily addressed. The stepped-wedge design would allow inference of benefit but would not be as helpful as the null seeking trial at defining the patients on whom to operate.

Sequential trials

Sequential trials are similar to the null seeking trial in that the results are consulted repeatedly as the trial is conducted but they are not made public, and so they do not influence the type of patient that is randomized. The sample size is not fixed at the beginning of the trial. Instead, stopping rules are used to determine the trial size on the basis of data so far collected. Sequential trials have major advantages over conventional binary trials in terms of ethics, economics and administration. Early in the evolution of sequential trial design, Armitage suggested very frequent interim analysis and O'Quigley's continual reassessment method examined the results one at a time and used them to alter the intervention (a drug dose) to achieve a specific effect in phase I trials[38]. This method was related to the null seeking trial but it was the intervention that was changed, not the patients who were randomized. Most sequential trials are further from the null seeking trial with less frequent interim analysis. They are grouped into those using sequential probability ratio tests and those using α-spending functions[39, 40]. Both types aim to reach a conclusion of a significant difference, or otherwise, between groups with a minimum of time, subjects and cost. They are an optimization of binary trials rather than a type of null seeking trial.

Futility analysis

Futility analysis is a method used to monitor the progress of clinical trials and to recommend stopping if it becomes clear that the likelihood of reaching a significant result by the end is low[41]. It is based on power calculations that assume the trends and features of data available at the interim analysis will be continued throughout the trial. Power calculated in this way is conditional on these assumptions being correct and is called conditional power.

Futility analysis is often used in conjunction with sequential trial designs. It is part of binary trial design and is aimed at assessing the likelihood of finding a significant difference between groups, not the location of a point of no difference. This means it cannot be readily adapted to fit the null seeking trial model.

Adaptive trials

Adaptive trial design uses the results of interim analysis to change features of the trial's design, such as the interval between interim analyses, the probability of randomization to either group, the sample size, or even the primary outcome measure. They are a development from sequential design where the points of interim analysis must be pre-specified. Outcome data are divided into blocks between sequential points of interim or final analysis. Combination functions, or conditional error functions to which they are equivalent, are used to inform design changes and to analyze the final result. These are functions of the data blocks and in some cases allow that the block size may be dependent on the data in previous blocks. All such functions assume that the data within each block are independent of those within all other blocks[42]. It is this last assumption that makes the combination functions of adaptive trials inapplicable to the null seeking trial, which uses the emerging result to influence the type of patient being randomized so that the emerging data themselves are dependant on previously collected data.

Patients' preference trials

When patients in a conventional binary trial are asked for their consent to participate, they understand that participation means being randomized to one treatment or another and that the outcome will be assessed. If they do not wish to be randomized they do not participate in the trial at all. Problems can arise with patients who have a preference for one or other of the treatments being compared. In some cases an interaction between a patient's preference for a treatment and their outcome following the treatment has been found. This particularly occurs in trials where the outcome measure is dependent on subjective reporting by the patient such as quality of life measures and visual analogue pain scales. In general, if a patient has the treatment they prefer their outcome will be better than if they have the treatment they did not prefer or if they are indifferent. On the other hand if a patient gets the treatment they did not want, they may report a poorer outcome or develop what is called *resentful demoralisation* because they did not get the treatment they wanted, leading to poor compliance with the trial and possibly withdrawing altogether.

The problem that this leads to is that rather than comparing the effect of one treatment with that of the other, the trial will now compare the effect of one treatment *plus its interaction with patient preference* with the effect of the other *plus its interaction with patient preference*. As this depends on the preferences of patients recruited, the result cannot necessarily be extrapolated to the rest of the patient population, which will make it difficult to apply to practice later.

Patient preference can affect trials in two ways. If patients have the option of not participating in the trial and receiving their chosen treatment, they are likely to do this rather than risk being randomized to a treatment that they don't want. This means that in cases where both treatments are available to patients outside of the trial, patients with a preference for one treatment or the other will be selected not to participate, and the result of the trial will only be directly applicable to patients who do not have a preference. In many cases the treatments being compared are not both available outside of the trial, in which case patients who have a preference for a treatment that is only available in the trial are likely to participate because there is a chance that they will be randomized to the treatment they want. If that happens, then their outcome due to the treatment and its interaction with their preference is likely to be better than the effect of the treatment alone. If it does not happen, then the outcome due to the treatment and its interaction with their preference is likely to be worse than that due to the treatment alone. They may even drop out of the trial altogether.

Several measures have been taken to reduce this problem, not all of them good. They all involve recording every patient's preference or indifference when they are recruited. The best approach is to conduct a trial as normal but to record patient preferences so that a secondary analysis can be undertaken to look for any interaction between their preferences and their treatment. If such an interaction is found, the trial result can be tailored according to individual patient's preferences and more reasonably applied[43].

Using patient preference as a secondary analysis tool like this does not prevent the problem of patients with preferences not participating in the trial at all, which means the result, when it comes, is not applicable to those with preferences. To partly deal with the problem, patients can be given three options – to choose the treatment they want or to be randomized. Both options will be within the research project and will involve measuring and recording the outcome, as is done in a null seeking trial. There are then randomized and non-randomized cohorts. From the points of view of ethics and recruitment, this is an improvement because patients can still participate even if they do not wish to submit to the main imposition of a randomized trial, that is to take an important clinical decision out of the hands of the patient and doctors. Additional data to those collected in a conventional trial will be available. As with a conventional trial they allow comparison of the two treatments being studied with unbiased randomize data but substantial data will also be obtained on treatment performances in the non-randomized patients. These data will be closely comparable to the randomized data in that they are collected in the same context, by the same people, with the same timing and the same outcome measurement instruments. This makes it possible to compare the randomized results with the non-randomized results, giving a measure of the degree of allocation bias occurring in the non randomized groups due to unknown causes. In a conventional trial design, these additional data are basically thrown away. While they do not have the status of randomized data, they are certainly not valueless!

Non-randomized cohorts also have disadvantages. The only unbiased comparison is with intention to treat analysis of the randomized data. With respect to this component of the trial, patient preference designs may detract from conventional binary trials because the number of patients randomized may be smaller than with conventional trials. The incentive for patients to participate in randomization is diluted. Patients participate in trials because they want to help other people. With the patient preference design they can participate without being randomized so it is likely that some patients who wish to participate in research, and who would agree to randomization if it were the only option, do not agree because of other options that still allow participation.

The worst approach is to recruit patients with preferences and randomize them on the understanding that they can reject the result of randomization if they don't like it and choose the treatment they want anyway. This only encourages crossovers and aggravates the hypothesis problem discussed in Chapter 2.

From the patient's point of view, the only difference between a patient preference trial and a null seeking trial is that only the latter gives them access to the results of previous participants. While this difference will probably be of some significance to patients, from a scientific point of view it is profound.

Chapter 13

Miscellaneous Features and Limitations

Blinding, crossovers and the hypothesis problem affect null seeking trials in different ways. The selection bias of treatment received and per protocol analysis mean that were they used, the null point found would not be the true null point. It would be the point at which the difference between the treatments plus the bias was nought. Analysis would still have to be by intention to treat and the hypothesis problem would still arise, but it could be lessened. A component of the hypothesis problem in binary trials is likely to arise from the shift in opinion that occurs in the light of the trial result. This is avoided in a null seeking trial because the shift in opinion occurs during the trial and does not happen abruptly at its end.

The null seeking trial may improve trial recruitment because it does not interfere with the decision-making process unless the decision is difficult. Furthermore, once the trial is under way, there is a real possibility that participating in it would benefit the patients themselves, rather than being of benefit only to future patients. This is because each patient could access accurate information on which to base their own decision.

The main uncertainty in the feasibility of the null seeking trial is the relationship between the evolving result, which is a numerical or graphical display, and the zone of equipoise which is a clinical entity. This would only become clear if a null seeking trial was conducted. Simulations with groups of surgeons suggests that the feedback loop would work as expected, but real-life clinical decisions are very different from hypothetical simulated scenarios without a real patient present.

The main limitation of the null seeking trial concerns its purpose. It is not meant to show a difference between treatments or allow causal inference from treatment to effect. Rather it starts from untestable assumptions about when treatments should be used, and provides an unbiased guide to optimal decision making. Under certain circumstances, such as if the trial were large or the null point moved significantly, a causal inference could be made, but it would be a secondary rather than a primary result. The design of null seeking trials therefore does not stand alone as a generator of causal inference. It only has meaning in the context of the generality of published data, prevailing beliefs and personal experience.

The ultimate objective of clinical research is to improve practice. The current focus of research is to demonstrate the effects of treatments and to assume that the influence on practice will follow if the demonstration is widely accepted. The null seeking trial falls down in this respect because it is not intended to demonstrate the effect of treatments. It does, however, have the potential to improve practice through a different mechanism. Rather than showing a benefit or allowing causal inference, its design has the ability to change practice for the better. The role of treatments under study would be defined in an unbiased way, establishing whether they were almost always beneficial, always useless, or somewhere in between. Decision-making would be informed by randomized data. Clinically relevant questions in surgery are frequently not of the form: *Will this operation improve the average outcome from this condition?* but *On which patients should this operation be used?* Useless operations tested with a null seeking trial would be abandoned, even though their uselessness may never be proved.

The potential utility of the null seeking trial can be illustrated by the case of appendicectomy. This is a commonly conducted, possibly life-saving operation. Hundreds of thousands of appendicectomy operations are done around the world every year. It has features that make it suitable for investigation with a null seeking trial. Appendicitis presents with abdominal pain. Some cases of abdominal pain are widely believed to need the operation, others are widely believed not to. Between these extremes lie intermediate cases forming a spectrum. Several numerical scales have been developed to guide the decision of whether or not to operate. Numerical outcome measures to quantify the effect of treatment are available in the relatively short term. Time to resolution of symptoms and the rate of symptom recurrence within the first year are among them. It would be a straightforward matter to establish a null seeking trial of appendicectomy, and the large volume of surgery done for appendicitis would mean that reliable results would become available quickly.

Compare this with the binary trial situation. Three small trials have been conducted on selected patient groups to assess the value of appendicectomy [44–46]. Their results were mixed with similar rates of resolution with and without surgery, but somewhat higher rates of symptom recurrence within a year without surgery. Trial data on appendicectomy are so scarce because of a lack of equipoise about whether the operation should be done or not. The issue of how to do the operation is less important than the issue of whether to do the operation and numerous trials have been conducted that compare different techniques, because equipoise is more widespread. The lack of randomized data on the value of appendicectomy, despite its great clinical and economic importance, argues for an alternative strategy to binary trials.

There are numerous other examples of decisions made in large numbers of cases, supported by minimal amounts of randomised data or none at all, that would lend themselves to investigation with null seeking trials. They extend beyond health care to include issues such as educational methods, road design, marketing, economics, ergonomics and industrial safety. Decisions that can expressed as numerical spectra, ranging from consensus one way through uncertainty to consensus the other, and with outcomes that can also be measured numerically, could be assisted in this way. For the present though, the concept remains theoretical, with unproven feasibility let alone utility.

Bibliography

1. Majeed, A.W., et al., *Randomised, prospective, single-blind comparison of laparoscopic versus small-incision cholecystectomy.* Lancet, 1996. **347**(9007): p. 989-94.
2. McMahon, A.J., et al., *Laparoscopic versus minilaparotomy cholecystectomy: a randomised trial.* Lancet, 1994. **343**(8890): p. 135-8.
3. Flum, D.R., *Interpreting surgical trials with subjective outcomes: avoiding UnSPORTsmanlike conduct.* JAMA, 2006. **296**(20): p. 2483-5.
4. Koutsourelakis, I., et al., *Randomised trial of nasal surgery for fixed nasal obstruction in obstructive sleep apnoea.* Eur Respir J, 2008. **31**(1): p. 110-7.
5. Rothstein, R., et al., *Endoscopic full-thickness plication for the treatment of gastroesophageal reflux disease: A randomized, sham-controlled trial.* Gastroenterology, 2006. **131**(3): p. 704-12.
6. Leclaire, R., et al., *Radiofrequency facet joint denervation in the treatment of low back pain: a placebo-controlled clinical trial to assess efficacy.* Spine, 2001. **26**(13): p. 1411-6; discussion 1417.
7. Weinstein, J.N., et al., *Surgical vs nonoperative treatment for lumbar disk herniation: the Spine Patient Outcomes Research Trial (SPORT) observational cohort.* JAMA, 2006. **296**(20): p. 2451-9.
8. Mendelow, A.D., et al., *Early surgery versus initial conservative treatment in patients with spontaneous supratentorial intracerebral haematomas in the International*

Surgical Trial in Intracerebral Haemorrhage (STICH): a randomised trial. Lancet, 2005. **365**(9457): p. 387-97.

9. Berger, J.S., et al., *Aspirin for the primary prevention of cardiovascular events in women and men: a sex-specific meta-analysis of randomized controlled trials.* JAMA, 2006. 295(3): p. 306-13.
10. Brox, J.I., et al., *Lumbar instrumented fusion compared with cognitive intervention and exercises in patients with chronic back pain after previous surgery for disc herniation: a prospective randomized controlled study.* Pain, 2006. **122**(1-2): p. 145-55.
11. Brox, J.I., et al., *Randomized clinical trial of lumbar instrumented fusion and cognitive intervention and exercises in patients with chronic low back pain and disc degeneration.* Spine, 2003. **28**(17): p. 1913-21.
12. Fairbank, J., et al., *Randomised controlled trial to compare surgical stabilisation of the lumbar spine with an intensive rehabilitation programme for patients with chronic low back pain: the MRC spine stabilisation trial.* Bmj, 2005. **330**(7502): p. 1233.
13. Fritzell, P., et al., *2001 Volvo Award Winner in Clinical Studies: Lumbar fusion versus nonsurgical treatment for chronic low back pain: a multicenter randomized controlled trial from the Swedish Lumbar Spine Study Group.* Spine, 2001. **26**(23): p. 2521-32; discussion 2532-4.
14. Mirza, S.K. and R.A. Deyo, *Systematic review of randomized trials comparing lumbar fusion surgery to nonoperative care for treatment of chronic back pain.* Spine, 2007. **32**(7): p. 816-23.
15. *Beneficial effect of carotid endarterectomy in symptomatic patients with high-grade carotid stenosis. North American Symptomatic Carotid Endarterectomy Trial Collaborators.* N Engl J Med, 1991. **325**(7): p. 445-53.
16. *Randomised trial of endarterectomy for recently symptomatic carotid stenosis: final results of the MRC European Carotid Surgery Trial (ECST).* Lancet, 1998. **351**(9113): p. 1379-87.
17. Helton, T.J., et al., *The optimal treatment of carotid atherosclerosis: a 2008 update and literature review.*

Postgrad Med, 2008. **120**(3): p. 103-12.

18. Kampmeier, R.H., *Final report on the "Tuskegee syphilis study".* South Med J, 1974. **67**(11): p. 1349-53.
19. Dowd, S.B. and B. Wilson, *Informed patient consent: a historical perspective.* Radiol Technol, 1995. **67**(2): p. 119-24.
20. Katz, R.V., et al., *The legacy of the Tuskegee Syphilis Study: assessing its impact on willingness to participate in biomedical studies.* J Health Care Poor Underserved, 2008. **19**(4): p. 1168-80.
21. Freedman, B., *Equipoise and the ethics of clinical research.* N Engl J Med, 1987. **317**(3): p. 141-5.
22. Gregson, B.A. and A.D. Mendelow, *International variations in surgical practice for spontaneous intracerebral hemorrhage.* Stroke, 2003. **34**(11): p. 2593-7.
23. Gibson, J.N. and G. Waddell, *Surgery for degenerative lumbar spondylosis: updated Cochrane Review.* Spine, 2005. **30**(20): p. 2312-20.
24. Mitchell, P. and J. Jakubowski, *Risk analysis of treatment of unruptured aneurysms.* J Neurol Neurosurg Psychiatry, 2000. **68**(5): p. 577-80.
25. Vindlacheruvu, R.R., A.D. Mendelow, and P. Mitchell, *Risk-benefit analysis of the treatment of unruptured intracranial aneurysms.* J Neurol Neurosurg Psychiatry, 2005. **76**(2): p. 234-9.
26. Mitchell, P., D.W. Ellison, and A.D. Mendelow, *Surgery for malignant gliomas: mechanistic reasoning and slippery statistics.* Lancet Neurol, 2005. **4**(7): p. 413-22.
27. Bowman, A.W. and A. Azzalini, *Applied Smoothing Techniques for Data Analysis*. Oxford Satstistical Science Series. 1997, Oxford: Clarendon Press.
28. Cleveland, W.S., *Robust Locally Weighted Regression and Smoothing Scatterplots.* Journal of the American Statistical Association, 1979. **74**(368): p. 829-836.
29. Cleveland, W.S., G. E., and S.M. W., *Local regression models*, in *Statistical models in S*, J.M. Chambers and T.J. Hastie, Editors. 1992, Kluwer Academic Publishers. p. 309 – 376.

30. Rice, J., *Bandwidth choice for nonparametric regression* Ann. Statist. Assoc., 1984. **12**: p. 1215-30.
31. Gasser, T., L. Sroka, and C. Jennen-Steinmetz, *Residual variance and residual pattern in nonlinear regression.* Biometrika, 1986. **73**: p. 625-33.
32. Beran, R., *Nonparametric regression with randomly censored survival data, Technical report.* 1981, University of California, Berkely.
33. Doksum, K.A. and B.S. Yandell. *Properties of regression estimates based uncensored survival data.* in *A Festschrift for Eric Lehmann.* 1982. Belmont California: Wadsworth International group.
34. Gentleman, R. and J. Crowley, *Graphical methods were censored data.* J. Amer. Statist. Assoc., 1991. **86**: p. 678-83.
35. Hardle, W. and B.A. W., *Bootstrapping in non parametric regression: local adaptive smoothing and confidence bands.* J. Amer. Statist. Assoc., 1988. **83**(401): p. 102-10.
36. Brown, C.A. and R.J. Lilford, *The stepped wedge trial design: a systematic review.* BMC Med Res Methodol, 2006. **6**: p. 54.
37. *The Gambia Hepatitis Intervention Study. The Gambia Hepatitis Study Group.* Cancer Res, 1987. **47**(21): p. 5782-7.
38. O'Quigley, J., M. Pepe, and L. Fisher, *Continual reassessment method: a practical design for phase 1 clinical trials in cancer.* Biometrics, 1990. **46**(1): p. 33-48.
39. Lan, K.K.M. and D.L. DeMetz, *Discrete sequential boundaries for clinical trials.* Biometrika, 1983. **70**: p. 659-663.
40. Todd, S., *A 25-year review of sequential methodology in clinical studies.* Stat Med, 2007. **26**(2): p. 237-52.
41. Lachin, J.M., *A review of methods for futility stopping based on conditional power.* Stat Med, 2005. **24**(18): p. 2747-64.
42. Schafer, H., N. Timmesfeld, and H.H. Muller, *An overview of statistical approaches for adaptive designs and design modifications.* Biom J, 2006. **48**(4): p. 507-20.
43. Manca, A., et al., *Randomized trial of two physiotherapy interventions for primary care back and neck pain patients: cost effectiveness analysis.* Rheumatology (Oxford), 2007.

46(9): p. 1495-501.

44. Roumen, R.M., et al., *Randomized clinical trial evaluating elective laparoscopic appendicectomy for chronic right lower-quadrant pain.* Br J Surg, 2008. **95**(2): p. 169-74.
45. Styrud, J., et al., *Appendectomy versus antibiotic treatment in acute appendicitis. a prospective multicenter randomized controlled trial.* World J Surg, 2006. **30**(6): p. 1033-7.
46. Eriksson, S. and L. Granstrom, *Randomized controlled trial of appendicectomy versus antibiotic therapy for acute appendicitis.* Br J Surg, 1995. **82**(2): p. 166-9.

Index

Made in the USA
Monee, IL
04 November 2026

67386531R00075